RF Measurements of Die and Packages

For a listing of recent titles in the *Artech House Microwave Library*, turn to the back of this book.

RF Measurements of Die and Packages

Scott A. Wartenberg

Artech House
Boston • London
www.artechhouse.com

Library of Congress Cataloging-in-Publication Data
Wartenberg, Scott A.
    RF measurements of die and packages / Scott A. Wartenberg.
        p.   cm. — (Artech House microwave library)
    Includes bibliographical references and index.
    ISBN 1-58053-273-x (alk. paper)
        1. Very high speed integrated circuits—Measurements.   2. Integrated circuits
    —Masks—Design and construction.   3. Microelectronic packaging.   4. Radio
    measurements.   I. Title.
    TK7874.7 .W37   2002
    621.3815—dc21   2002016431

British Library Cataloguing in Publication Data
Wartenberg, Scott A.
    RF measurements of die and packages. — (Artech House microwave library)
        1. Integrated circuits—Testing   2. Radio circuits—Testing   3. Probes
        (Electronic instruments)
    I. Title
    621.3'815

    ISBN 1-58053-273-x

Cover design by Yekaterina Ratner

© 2002 ARTECH HOUSE, INC.
685 Canton Street
Norwood, MA 02062

International Standard Book Number: 1-58053-273-x
Library of Congress Catalog Card Number: 2002016431

10 9 8 7 6 5 4 3 2 1

Contents

Preface

In the early days, S-parameters were measured with a six-port reflectometer constructed entirely of components found in the lab [1]. In assembling the reflectometer, the engineer strove to locate connectors that produced little reflection and couplers with excellent directivity. The reflectometer's parasitic effects were de-embedded from the final measurement using pencil and paper. Because of the reflectometer's custom design, accurate and repeatable measurements of extremely large and small impedances were a challenge.

The advent of the automated *vector network analyzer* (VNA) alleviated the bulk of these problems. At the same time, a reliable VNA provided for the development of accurate RF probing tools. This book explains how to use coplanar probes and test fixtures to characterize *radio-frequency integrated circuits* (RFICs) and *monolithic microwave integrated circuits* (MMICs).

The growth of the RF wireless market along with digital integrated circuits of clock speeds >1 GHz have created a solid need for this introductory book. To correct for nonidealities in the test system, both coplanar probes and test fixtures must be calibrated (Chapter 2). RF on-wafer probes are explored in detail, both coplanar probes used in the lab (Chapter 3) and membrane probes used in high-volume testing (Chapter 4). Chapter 5 covers test fixtures, the alternative approach to on-wafer probing. The most intricate task for coplanar probes, on-wafer device characterization, is described in Chapter 6, while a description of the complete RF test system is given in

Chapter 7. Since most die are sold as packaged components, Chapter 8 provides an understanding of how to RF-characterize the die's package.

The reader is encouraged to send corrections, criticisms, and suggestions to either Artech House or myself. They will be incorporated into future editions. Thanks for your interest.

Reference

[1]   Engen, G., "A Review of the Six-Port Measurement Technique," *IEEE Microwave Symposium Digest*, 1997, pp. 1171–1172.

Acknowledgments

I thank (in alphabetical order) Terry Burcham, Domingo Figueredo, Reed Gleason, Steve Hamilton, Troels Kolding, David Reid, Jeff Sinsky, Patrick Tang, and the staff at Artech House for their expert suggestions in preparing this manuscript. My utmost appreciation goes to my wife and three children, who deeply missed my company during evenings and weekends. This book is dedicated to them.

1

Introduction

Put simply, the goal of this book is to enable the reader to go into the lab and make better RF measurements. During the course of designing RFICs and MMICs, much revolves around the RF data. It is presented to the engineering staff to validate a computer-aided design. It is used to iterate *integrated circuit* (IC) mask layouts used by the foundry. It guides the designer's overall RF understanding. All of these directly impact the design's cycle time and cost, and shortening this cycle is the surest way to product success.

The beginning of this chapter introduces the topics covered in subsequent chapters. Chapters are ordered in a building-block fashion, with each chapter laying the foundation for the[1] next. Understanding the basic composition of an RF test system is the first building block. To this end, the second part of this chapter details the essential elements in an RF test system. The three fundamental elements are: the test system interface to the *device-under-test* (DUT) (either coplanar probes or a test fixture); the VNA; and the connecting cables.

1.1   Topics Covered in This Book

1.1.1   Calibration

To ensure the validity of the measured data, the test equipment must be calibrated. Calibration serves to quantify the degree of error in a test system. Chapter 2 explains the errors found in an RF test system and how they can be mathematically removed by applying error models. Different methods of RF calibration are available, each based on measuring a set of standards in place of the DUT. At the end of Chapter 2, a list of common tips and tricks helps the reader put into practice the calibration concepts presented.

1.1.2   Coplanar Probes

Physically contacting the DUT, the coplanar probe becomes the final link in the RF test system to the wafer. Chapter 3 discusses the construction and RF characteristics of coplanar probes, beginning with coplanar waveguide theory. Training in their proper use develops habits that enhance the probe's accuracy and lifetime. The configuration of the coplanar probe depends on the die's layout of its probe pads, as well as the die's RF function. The coplanar probe has its own unique set of calibration standards. Chapter 3 examines the electrical issues to consider when using and designing these standards.

Another style of RF probe, the membrane probe, is well suited for high-volume testing and is discussed in Chapter 4.

1.1.3   High-Volume Probing

High-volume RF testing is the domain of manufacturing assembly. Issues surrounding high-volume testing are distinct from those in the lab, where individual die are manually probed one at a time. Test speed is the central issue driving the design of manufacturing test systems and, to some degree, the design of the product. Chapter 4 describes a typical high-volume test setup, focusing on the membrane probe. Analogous to the coplanar probe, the membrane probe is the RF interface from the test system to the DUT. Because of the speed of the die moving through the RF test system, unique issues arise.

1.1.4   Test Fixtures

Typically the final product is not a bare die but one that has been wire-bonded and packaged. For packaged die, the best RF interface to the test system is a test fixture [1] (see Figure 1.1). The ideal test fixture has [2] the following characteristics:

- No loss and no electrical length;
- Flat frequency response;
- Perfect impedance match between the test system and DUT ports;

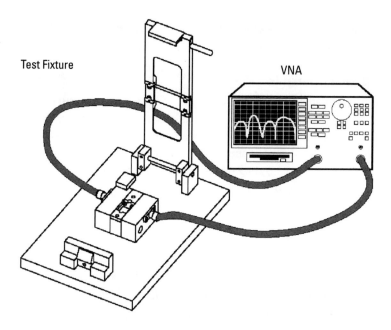

Figure 1.1 An RF test fixture and VNA for measuring packaged devices.

- No crosstalk between the fixture's input and output ports;
- Fast, easy, repeatable connections.

In practice, each of these points are met by the following:

- Keeping the fixture's amount of loss and phase errors << the system's measurement uncertainty;
- Making sure the DUT bandwidth << fixture bandwidth;
- The characteristic impedance Z_0 of the test system = the test fixture Z_0 = the DUT Z_0;
- The test fixture crosstalk << the DUT's loss or isolation;
- The measurement repeatability << the margin of the DUT test specifications.

Chapter 5 covers how to design an RF test fixture. In particular, it discusses the parasitic effects of the fixture on the DUT, as well as a detailed discussion of how to design RF transitions in the fixture. The often-challenging task of calibrating a test fixture is also included.

1.1.5 On-Wafer Characterization

Chapter 6 explains how to use RF coplanar probes to measure die on the wafer. The DUT is simply a thin film deposited on the wafer, too small to contact directly without disturbing it. Finding its RF behavior with coplanar probes requires probe pads and interconnecting lines. The discussion in Chapter 6 centers on how to de-embed the effects of the probe pads and interconnects leading to the DUT.

1.1.6 RF Test Systems

Moving beyond the DUT-to-coplanar probe interface, Chapter 7 describes the rest of the RF test system. The choice of test system depends on the RF quantity under study. Highlighted are three RF test systems and how to use them. These are a noise measurement system, a high-power RF test system, and a thermal (both hot and cold) measurement system.

1.1.7 Package Characterization

The package is an essential part of an RFIC or MMIC component. Its importance is often underestimated, neglected by the designer until the last moment. The package can have a demonstrable impact on the final product's RF performance. This chapter explores package testing, breaking the test fixture down section by section and explaining the RF aspects of its design. The chapter also covers techniques on how to characterize packages, both empty and with the die mounted inside. Calibration issues are discussed. The chapter ends by explaining how to characterize popular package styles.

1.2   Components of an RF Test System

The typical RF test system is composed of a VNA, RF cables, bias cables, and the interface to the DUT (see Figure 1.2). The interface is either RF on-wafer probes or a test fixture. The quality of the test system's RF performance depends on the VNA, the reliability of the RF cable and fixture connections, and the calibration quality.

Attempts at RF measurement fail when parts of the test system are corrupted by errors. At worst, these sources of error impede making any measurements. At best, the errors degrade RF measurement quality. This section details the components of an RF test system, highlighting where degradations may be found in the test setup and how to correct for them.

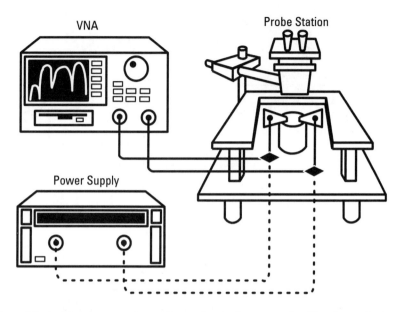

Figure 1.2 An RF test system for probing active devices on-wafer. The diamonds denote the locations of the bias tees.

1.2.1 VNA

Figure 1.3 shows the internal schematic of a VNA. An RF source drives either port 1 (P1) or port 2 (P2) of the DUT while the opposite port path is switched to the load R_L. The VNA is constructed with either three (a_0, b_1, b_2) or four (a_1, a_2, b_1, b_2) signal samplers.

In a three-sampler VNA, the switching between forward and reverse paths results in a different load being presented to P1 and P2. Yet the calibration equations assume the two paths have equal terminations. In reality, the two R_L loads are not identical, a source of error.

When the RF signal is down-converted to an *intermediate frequency* (IF), harmonics arise. To avoid mixing products, some VNAs allow the user to set the IF filter bandwidth. Choosing a narrow IF bandwidth results in less noise. However, phase-lock between the RF source and *local oscillator* (LO) can be lost, skipping to the next harmonic (see Figure 1.4). In general, a narrow IF bandwidth lowers the noise floor, enabling the measurement of smaller signals. The improved sensitivity yields a lower measurement noise floor, a gauge of the smallest signal that can be measured.

The perfect VNA has infinite isolation between ports, infinite directivity, no impedance mismatches anywhere, and a flat frequency response. In

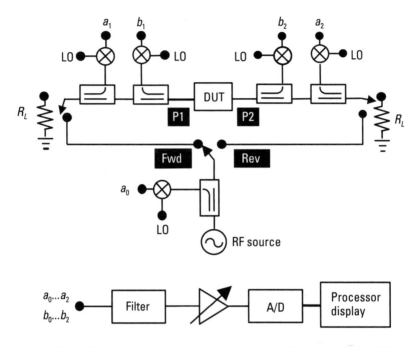

Figure 1.3 Block diagram of the vector network analyzer. A three-sampler VNA uses one coupler (a_0) to detect the incident signal, while a four-sampler VNA uses two couplers (a_1 and a_2). Reflected signals are detected using b_1 and b_2.

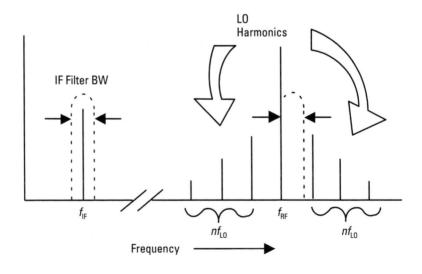

Figure 1.4 Harmonic mixing of the RF and LO sources inside the VNA. The dotted line shows the IF filter bandwidth (BW), while nf_{LO} are the harmonics around f_{RF}.

reality, the VNA's internal components are nonideal. To better understand the sources of error in the VNA, some key terms must be defined. *Directivity* is a measure of how well the VNA's couplers separate the forward and reflected waves. A high value of directivity is desirable. Poor directivity allows leakage of the signal into the coupled path. *Source match* is how closely the source path's R_L matches 50Ω, while *load match* is the quality of the load path's R_L termination. The two matches can be impacted by discontinuities between the RF source and the sampling paths. *Frequency response tracking* is how well the magnitude and phase of a signal passing through the VNA compares to (or tracks) the reference signal. This reveals the frequency response of the RF source and the internal samplers. *Transmission mismatch* is due to impedance mismatch between the test system's ports and the DUT's ports. *Isolation* is the leakage of energy between the test and reference channels inside the VNA. Leakage between transmitted and reflected paths within the VNA contaminates low-level signals.

Proper operation of the VNA by the user is the easiest way to improve RF measurements. For instance, some VNAs allow the user to choose among the frequency sweep modes. The *step sweep* mode phase-locks the LO and RF synthesizers at each discrete frequency point, providing better repeatability. The *ramp sweep* mode phase-locks the LO and RF synthesizers at the first frequency point only. Further points depart from the initial phase-lock, causing phase drift to accumulate. Ramp mode has a faster test time but less measurement accuracy.

When the VNA processor display is set to a fine-enough scale, the once-flat trace shows fluctuations and spikes with frequency. A good example is the insertion loss measurement of a 50-Ω transmission line. What appears as a straight line on a 10-dB per division log magnitude plot looks jagged on a 0.1-dB per division scale. The jaggedness is a result of thermal noise and the noise figure of the amplifiers within the VNA. Averaging several measurements using the VNA's average feature improves the measurement. Next, remove the 50-Ω transmission line and attach 50-Ω loads to P1 and P2. The log magnitude of the insertion loss will give the VNA's noise floor, usually −70 dB or less.

1.2.2 LCR Meter

Instead of a VNA, an *inductance capacitance resistance* (LCR) meter is often used to measure chip components such as capacitors, inductors, and resistors. Common ones operate at 1 MHz, contacting the DUT with needle probes. Measuring at high frequencies can reveal parasitic effects that do not appear at 1 MHz.

Aside from operating at a low-fixed frequency, LCR meters do not allow flexibility of circuit design. The equivalent circuit is always assumed to be an inductor-capacitor-resistor (LCR) combination. With a VNA, the measured S-parameters can be fit to any circuit configuration.

1.2.3   RF Cables

RF cables connect the test system to the DUT. Quality RF cables exhibit minimal insertion loss and *voltage standing-wave ratio* (VSWR). To achieve this, the cable's dielectric core has a low dielectric constant and loss tangent. The outer diameter of the center conductor provides plenty of conductive surface, diminishing its inductance.

Two types of RF cabling, semirigid and flexible, are available. Semi-rigid cables have a solid aluminum or copper jacket over a semiflexible dielectric protecting a solid center conductor. In comparison, a flexible RF cable is a more complicated arrangement of plastic-encased, aluminum-braided shielding over a flexible dielectric that covers a stranded center conductor. In general, good RF cables have a flexible outer skin, are thin, and have a smooth conductor surface.

Flexible cables are more expensive than semirigid, but the added expense provides advantages. Turning semirigid cable around corners requires a miniature pipe-bender that resembles a plumber's tool. The signal phase changes when the cable is bent because the outer radius becomes larger than the inner. Microscopic cracks can occur in the solder from the cable to the connector, resulting in an unreliable ground connection. Low-loss, semirigid cabling allows more dynamic range for the VNA.

In both semirigid and semiflexible, bent cables will dent the dielectric, forming an impedance mismatch at the dent. Such mismatches present a discontinuity similar to connector adapters. Flexing a cable degrades S-parameter repeatability to no better than −40 dB (or less than 1% of the transmitted signal). Physical motion such as raising and lowering the probes off the wafer's surface flexes cables. The solution is to fix the cables to the probe station and move the wafer chuck up and down, resulting in repeatability better than −60 dB. The metal-to-metal connection of the cable's center conductor and the connector pin can create a nonlinear interface, leading to intermodulation distortion in mixer applications.

The following are some general tips when using high-frequency cables [3]:

- Keep the number of bends to a minimum.
- Make the bend radii as large as possible.

- Operate in a temperature-controlled environment.
- Use the minimum number of cables, ideally connecting the DUT directly to the test port.
- Minimize movement.
- Keep the cable length short to avoid noise at high frequency [4].

In general, changes in the environment cause changes in the cable's phase delay. Small changes in the cable's phase length and insertion loss can occur with temperature. Some manufacturers offer pairs of phase-matched cables with identical electrical length and insertion loss.

1.2.4 Bias Cables

In the lab, coaxial cables are often used to supply bias to the DUT. Although they are easily found around the lab, using coaxial cables to connect the DUT to the dc power supply has disadvantages. When current flows, a voltage drop occurs along the length of the cable. Also, current leaks through the dielectric between the coaxial center conductor and its outer shielding.

Applications demanding precision current levels below 1nA require triaxial cables. A triaxial cable has three concentric conductors: force/sense, guard, and common. Typically, the innermost conductor is either the force or the sense, the next conductor guard, and the outermost common (see Figure 1.5). Voltage is applied on the force and measured with the sense. Within the power supply is a unity-gain amplifier that drives the same voltage on the guard surrounding the force. For low-current measurements, presenting the same potential on the force and guard negates leakage between them as well as capacitance. Lengthening the integration time of the dc measurement overcomes the cable's capacitance charging time. A high-quality dielectric in triaxial cables results in femtofarads of capacitance between force/sense and common.

With low resistance and negligible current, the triaxial sense line precisely reads the forced voltage, thereby capturing any voltage drop along the force line. At the power supply end of the sense line, a high impedance prevents current from flowing into the power supply.

The common outer conductor Faraday-shields the cable from external *electromagnetic interference* (EMI). Well-built bias tees should be enclosed in a metal box that connects to common. This completely shields the bias path end-to-end from power supply to DUT. To eliminate all EMI, encase the DUT in the same manner of shielding and connect to the same common

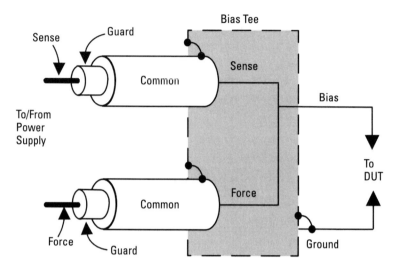

Figure 1.5 Force and sense triaxial cables. These connect to the DUT via a bias tee. In this configuration, the common also serves as ground.

(see Figure 7.2 in Chapter 7). With triaxial cables, the common is not necessarily ground. When the common is not ground, the DUT can connect to the power supply's ground using a separate cable.

Regarding its RF performance, the equivalent circuit of any coaxial or triaxial cable is a distributed series inductance with shunt capacitance. Long cables have large inductance, useful in blocking noise generated by a dc supply from getting into the RF channel. A practical advantage is that large inductance dampens amplifier oscillations. Conversely, high-inductance cables create undesirable ripple on the rising and falling edges of square-wave pulses. Solutions are either shorter cables with lower inductance or slowing the pulse's rise time.

1.2.5   Bias Tees

Active devices such as transistors and diodes require a voltage and current. Needle probes are commonly used to supply bias on-wafer. During an RF test, the problem becomes how to de-embed the RF effects of the needles. Not only do they have high inductance, needle probes will couple to other nearby probes. Attaching a bias tee to the input of an RF probe is a simpler way to supply bias to the DUT since the probe can simultaneously supply both RF and dc.

The bias tee is a three-port network. RF is fed into one port and dc into another, while RF and dc depart together from the third port. The dc input includes a lowpass network to keep the applied RF from disturbing the power supply and vice versa. Similarly, the RF input has a highpass network to keep dc from flowing into the VNA. Commercial bias tees are specified by their frequency range, RF power handling, and current rating.

1.3 RF Connectors

1.3.1 Connector Types

Common RF coaxial connector styles are 3.5 mm, *Subminiature A* (SMA), and *Amphenol Precision Connector–7 mm* (APC-7). The 3.5 mm and SMA are sexed while the APC-7 is not. The APC-7 is designed for the lowest *standing-wave ratio* (SWR) and has the best repeatability. Because they are small, the 3.5 mm and SMA have an advantage at higher frequencies. While the SMA and 3.5 mm appear identical, the SMA with its dielectric core operates to 24 GHz while the 3.5 mm with an air core extends to 34 GHz.

1.3.2 Making the Connection

Repeatability and accuracy of RF/microwave connectors critically depend on alignment of the mating surfaces. APC-7 connectors are prone to misalignment of the spring-loaded center conductors called collets. With 3.5 mm and SMA connectors, the male center conductor pin fits snugly into the female contact fingers. At the same time, the two connectors' dielectrics contact along a flat plane. To make sure this occurs properly, periodically use a connector gauge to adjust the center pin length, usually just a few ten-thousandths of an inch. Similarly for the APC-7, the collets extend just enough for the two planes to make contact when properly tightened. Using a torque wrench ensures the mating planes make uniform contact without overtightening and crushing them together.

An air gap in the mating plane can arise, either due to a gap in the dielectrics or because of a gap between center pins. An air gap between male and female connectors manifests itself in the return loss (see Section 2.2 in Chapter 2). A dielectric gap or pin gap can contribute to inaccurate coaxial calibration.

Different-size connectors will sometimes mate mechanically but not electrically. For instance, SMA, 3.5-mm, and 2.92-mm connectors will all mate mechanically, but the electrical interface will exhibit a capacitive mismatch.

There are two types of female connector pins, slotted (the most common) and slotless. To mechanically accept male pins of various diameters, the female connector pin often has a slot cut in its diameter. Slotless female pins are more inductive and expensive than their slotted counterparts. Yet slotless female pins are more repeatable since their diameter does not change with the size of the male pin. While mating mechanically, a change in the diameter of the male pin will contribute to reflections.

1.3.3 Connector Care

When mating two connectors, always tighten the nut rather than rotate the cable. The center conductors grind and wear while turning, leaving metal filings to dirty the interface. Such metal filings or particles originate from the connector threads. Dirt collects in the connector cavities. The connectors are cleaned using a blast of compressed air or a soft cloth or swab dampened with *isopropyl alcohol* (IPA). Do not dip the connector into the IPA as dirt will collect in its crevices as it dries.

When tightening two connectors together, hard dirt particles caught in between will plow scratches into the dielectric, manifested as carved concentric rings in the dielectric mating surfaces. In contrast, storing connectors facedown results in straight scratches across the dielectric faces.

1.4 RF Connector Adapters

Adapters transition between connector types and genders. Because adapters often involve transforming between diameters, their design is important. Quality adapters generate little reflections. Sometimes both test ports can have the same gender connector, complicating calibration. A DUT with the same gender connector on both ports is referred to as a noninsertible device. In this case, the contribution of the adapter to the calibration must be removed [5, 6].

1.5 The Probe Station

Properly understanding the role of the RF probe in the test system requires a description of a probe station. To begin, the wafer is placed on a tall column, or chuck, slightly larger than the diameter of the wafer. A vacuum line runs through the probe station to the center of the chuck, securing the wafer on top. Rather than move the probes from die to die, the chuck moves the wafer from die to die. Z-motion raises and lowers the chuck height with respect to the probes.

The probes are fixed to micropositioners secured to the probe station. The micropositioners are precision micrometers, enabling fine movement in the *x*, *y*, and *z* dimensions. A single micropositioner can hold multiple dc and RF probes. The RF probe is usually mounted to a micropositioner whose placement accuracy is within 1 micron. Placing the probes in the exact same spot on the wafer or on the calibration substrate each time is a challenge and can be a source of measurement error. As the chuck is stepped across the wafer, the probe tips can lose alignment with the die's pads. One way RF probe misplacement manifests itself is on the Smith chart. For instance, a 5-micron misplacement of the probes on the calibration standards can result in ±11 pH in pad inductance (see Figure 3.11 in Chapter 3). Such detailed alignment becomes more important as the frequency increases.

Minute vibrations can cause the probes to vibrate on the DUT's pads during the RF test. To prevent this, the entire probe station should sit on a vibration table. Bellows in the vibration table suspend the station on a cushion of air, dampening floor vibrations.

Depending on the application, probe stations are designed either for sample probing in the lab or for manufacturing. In the lab, chuck movement and DUT measurement are manual operations. Conversely, RF probing on the manufacturing line demands speed, where a computer automates wafer loading, chuck movement from die-to-die, and RF measurement.

1.6 Summary

This chapter serves as an introduction to the components of an RF test system. These components will be referred to again as concepts are further developed in Chapters 2 to 9.

References

[1]   Hewlett-Packard, "In-Fixture Measurements Using Vector Network Analyzers," *Hewlett-Packard Application Note AN,* 1997, pp. 1287–1289.

[2]   Hewlett-Packard, "Accurate Measurement of Packaged RF Devices," *RF and Microwave Device Test Seminar,* 1995, pp. 1–44.

[3]   Haag, I., "Flexible Cables Ease Wafer Test Probing to 65 GHz," *Microwaves & RF,* Vol. 31, No. 3, 1992, pp. 206–210.

[4] Jones, K., E. Strid, and K. Gleason, "mm-Wave Wafer Probes Span 0 to 50 GHz," *Microwave Journal*, Vol. 30, No. 4, 1987, pp. 177–183.

[5] Olney, D., "The Effect of Adapters on Vector Network Analyzer Calibrations," *Microwave Journal*, Vol. 36, No. 11, 1994, pp. 60–72.

[6] Oldfield, B., "The Connector Interface and Its Effect on Calibration Accuracy," *Microwave Journal*, Vol. 38, No. 3, 1996, pp. 106–114.

2

Calibration

Before an accurate measurement can be made, the test system must first be calibrated. Calibration de-embeds imperfections in the test system by measuring known quantities called standards. The RF characteristics of the standards are found beforehand. By measuring the standards mounted in place of the DUT, the test system's imperfections can be isolated, quantified, and mathematically removed.

This chapter examines RF calibration, how it is performed, the causes of calibration error, and where they are located in a test setup. A variety of calibration methods and standards are explored. Tips and tricks to help the reader perform expert high-frequency calibrations are presented at the end of the chapter. In many sections, examples are given based on using coplanar probes.

2.1 Test System Errors: Random or Systematic

An RF test system experiences two types of errors: random and systematic [1]. Random errors occur irregularly during the measurement. Varying unpredictably, they arise from semiconductor components within the test equipment. Examples include thermal excitation of carriers (Johnson noise) and minority carriers drifting across junctions (Schottky noise), both caused by changes in the room temperature. These random errors manifest themselves in the VNA display as low-level signals spuriously emerge from the noise floor. On the other hand, systematic errors are easier to identify but cover a broader range of phenomena. For instance, cables inherently exhibit inductance and bias tees capacitance at high frequencies. Also, dirty connectors make poor RF connections. When the RF error is measured repeatedly and predictably, it classifies as systematic error. In most test setups, both random and systematic errors exist at the same time. Of the two error types, systematic errors are more significant because they are easier to correct than random ones. De-embedding systematic errors is the primary focus in this chapter.

2.2   Concept of a Reference Plane

When engaged in a measurement, the RF test system and DUT together form a complete circuit. As the frequency increases, discriminating between the two becomes a delicate issue. The simplest method of separation is to sketch a line on a schematic, dividing the test system and the DUT. Subsequent measurements of the DUT are then referred to this dividing line. The DUT being measured is inside the line, while the test system is outside. This is the concept of a reference plane, delineating where the test system ends and the DUT begins.

There are two types of reference planes: physical and electrical. While drawing a line on a schematic and relating it to the physical setup is simple enough, the act of joining connectors reveals inherent complications (see Figure 2.1). Not properly tightening the nut leaves an air gap so that the reference planes do not meet. This physical gap serves as an unaccounted-for part of the circuit and a source of measurement error.

Distinct from the physical reference plane is the electrical reference plane. The act of calibration defines its location. When RF coupling occurs, defining the electrical reference planes can be difficult (see Figure 2.2). For instance, the male pin on the right capacitively couples to its own shield. When mated, this capacitance lies to the left of the female's electrical reference plane, adding to the response of the female connector. While it is valid to arbitrarily allot this parasitic capacitance to one connector or the other, such decisions should be clearly spelled out during calibration. Otherwise, it can add confusion to the measurement.

Upon calibration, the electrical calibration reference planes should align. At the same time, the physical reference planes can overlap. For instance, consider the physical reference plane at the edge of each connector

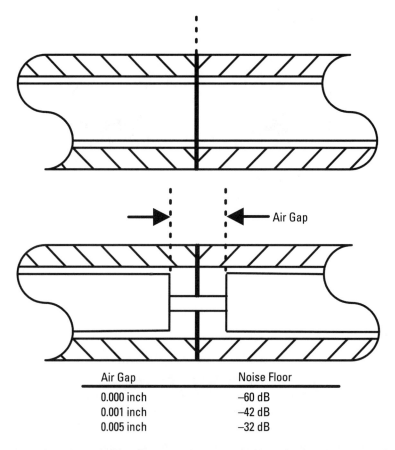

Figure 2.1 Connectors properly tightened result in the reference planes aligning (top). When loose, a small air gap arises between the reference planes (middle). At bottom is the effect on the noise floor at 26.5 GHz caused by an air gap in a 3.5-mm connector.

in Figure 2.2. When mated, the physical planes extend beyond one another into the body of the other connector.

2.3 Error Models

The goal of any calibration is to provide the DUT with an electrically pure connection to the test system's terminals. Thus the signal at the test ports should have zero magnitude, no phase shift, and a characteristic impedance

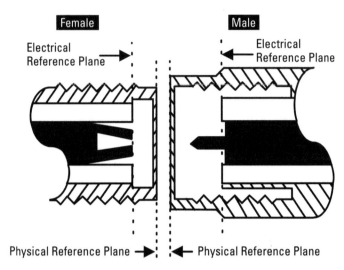

Figure 2.2 Coaxial connector interface, illustrating the difference between the electrical and physical reference planes.

Z_0. Yet every test setup has imperfections that corrupt the measurement. Mathematically placing an error model between the VNA and the DUT is one way to correct for these imperfections (see Figure 2.3). Specifically, an error model corrects for the directivity of internal couplers, impedance mismatches at the test system ports, the frequency response of the transmission and reflection paths, and crosstalk between the input and output paths [2].

Calibration quantifies the error model between the VNA test system and the DUT. Connecting standards at the output reference plane (ports 1 and 2) uncovers the terms in the error model. Finding each error model term requires multiple measurements. In general, error models are linear and assume one mode of propagation, the dominant mode.

A requirement for a valid error model is a static system configuration. In the VNA, an internal PIN diode switch alternates the RF source between forward and reverse directions with the inactive port terminated in R_L (see Figure 1.3). Since the impedance of the RF source is different from the load R_L, the system's forward and reverse responses behave differently. Hence, forward and reverse signal flow graphs are defined separately, depending on the switch setting.

Isolation is a distinctly separate component of the error model, encompassing coupling between the two paths. An example of poor isolation is when radiation or coupling occurs between the test ports. If the isolation

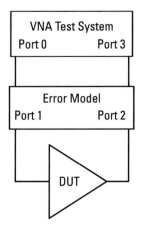

Figure 2.3 Functional diagram of the error model, mathematically correcting for test system imperfections.

amount changes when another DUT is inserted, then it is not de-embedded but instead contributes to the DUT measurement. The isolation calibration step should be done carefully, since a poor one can lead to unphysical measurement values.

2.3.1 Signal Flow Graph

One way to implement an error model is with a signal flow graph. In a two-port measurement, there are separate forward and reverse signal flow graphs. Figure 2.4 shows the forward signal flow graph. A single signal flow graph is composed of six terms (directivity, source match, reflection tracking, load match, transmission tracking, and isolation), each representing a unique source of error. To find each term, mount calibration standards one-at-a-time at test port 1. Afterwards, the information from the six calibration measurements is used to simultaneously solve six equations for six terms. Figure 2.5 shows the six systematic errors in the reverse direction. For instance, measuring the port 2 return loss helps to find the reflection terms E_{DR}, E_{SR}, and E_{RR}.

Connecting a thru-between the two ports will give the forward and reverse transmission terms E_{LF}, E_{LR}, E_{TF}, and E_{TR}. These error terms assess the frequency response of the transmission signal path and the impedance of the ports. An isolation measurement will reveal the crosstalk error terms E_{XF} and E_{XR}. Because they are often smaller than the connector repeatability error, the isolation terms are regularly neglected.

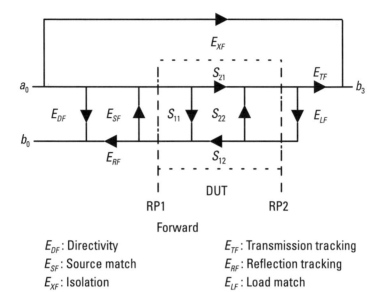

E_{DF}: Directivity E_{TF}: Transmission tracking
E_{SF}: Source match E_{RF}: Reflection tracking
E_{XF}: Isolation E_{LF}: Load match

Figure 2.4 Signal flow graph in the forward direction. The dotted line defines the DUT, while RP1 and RP2 are the calibration planes.

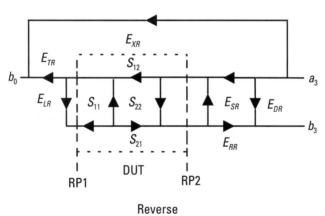

E_{DF}: Directivity E_{TF}: Transmission tracking
E_{SF}: Source match E_{RF}: Reflection tracking
E_{XF}: Isolation E_{LF}: Load match

Figure 2.5 Signal flow graph in the reverse direction. The dotted line defines the DUT, while RP1 and RP2 are the calibration planes.

For reference, the calibration equations (2.1)–(2.4) used to calculate the DUT's S-parameters ($S_{11,\text{DUT}}\ldots S_{22,\text{DUT}}$) from the measured S-parameters ($S_{11M}\ldots S_{22M}$) are

$$
S_{11,\text{DUT}} = \frac{\left\{\left(\dfrac{S_{11M}-E_{DF}}{E_{RF}}\right)\left[1+\left(\dfrac{S_{22M}-E_{DR}}{E_{RR}}\right)E_{SR}\right]\right\}-\left[\left(\dfrac{S_{21M}-E_{XF}}{E_{TF}}\right)\left(\dfrac{S_{12M}-E_{XR}}{E_{TR}}\right)E_{LF}\right]}{\left[1+\left(\dfrac{S_{11M}-E_{DF}}{E_{RF}}\right)E_{SF}\right]\left[1+\left(\dfrac{S_{22M}-E_{DR}}{E_{RR}}\right)E_{SR}\right]-\left[\left(\dfrac{S_{21M}-E_{XF}}{E_{TF}}\right)\left(\dfrac{S_{12M}-E_{XR}}{E_{TR}}\right)E_{LF}E_{LR}\right]}
$$

(2.1)

$$
S_{21,\text{DUT}} = \frac{\left[1+\left(\dfrac{S_{22M}-E_{DR}}{E_{RR}}\right)(E_{SR}-E_{LR})\right]\left(\dfrac{S_{21M}-E_{XF}}{E_{TF}}\right)}{\left[1+\left(\dfrac{S_{11M}-E_{DF}}{E_{RF}}\right)E_{SF}\right]\left[1+\left(\dfrac{S_{22M}-E_{DR}}{E_{RR}}\right)E_{SR}\right]-\left[\left(\dfrac{S_{21M}-E_{XF}}{E_{TF}}\right)\left(\dfrac{S_{12M}-E_{XR}}{E_{TR}}\right)E_{LF}E_{LR}\right]}
$$

(2.2)

$$
S_{12,\text{DUT}} = \frac{\left[1+\left(\dfrac{S_{11M}-E_{DR}}{E_{RR}}\right)(E_{SR}-E_{LR})\right]\left(\dfrac{S_{21M}-E_{XR}}{E_{TR}}\right)}{\left[1+\left(\dfrac{S_{11M}-E_{DF}}{E_{RF}}\right)E_{SF}\right]\left[1+\left(\dfrac{S_{22M}-E_{DR}}{E_{RR}}\right)E_{SR}\right]-\left[\left(\dfrac{S_{21M}-E_{XF}}{E_{TF}}\right)\left(\dfrac{S_{12M}-E_{XR}}{E_{TR}}\right)E_{LF}E_{LR}\right]}
$$

(2.3)

$$
S_{22,\text{DUT}} = \frac{\left\{\left(\dfrac{S_{22M}-E_{DR}}{E_{RR}}\right)\left[1+\left(\dfrac{S_{11M}-E_{DR}}{E_{RF}}\right)E_{SF}\right]\right\}-\left[\left(\dfrac{S_{21M}-E_{XF}}{E_{TF}}\right)\left(\dfrac{S_{12M}-E_{XR}}{E_{TR}}\right)E_{LR}\right]}{\left[1+\left(\dfrac{S_{11M}-E_{DF}}{E_{RF}}\right)E_{SF}\right]\left[1+\left(\dfrac{S_{22M}-E_{DR}}{E_{RR}}\right)E_{SR}\right]-\left[\left(\dfrac{S_{21M}-E_{XF}}{E_{TF}}\right)\left(\dfrac{S_{12M}-E_{XR}}{E_{TR}}\right)E_{LF}E_{LR}\right]}
$$

(2.4)

2.3.2   Error Adapter

Rather than going to the trouble of separating out each individual error path shown in Figures 2.4 and 2.5, all the error terms can be lumped together into a single error adapter. Viewed as a mathematical adapter between the test system and the DUT, the error adapter has eight terms e_{mn}, four for each port (see Figure 2.6). When the forward and reverse paths couple, it adds eight leakage terms (see Figure 2.7). Seven of the eight terms in Figure 2.6 are independent. That is, seven transmission-reflection measurements are all that are necessary to find the eight terms, although eight measurements are usually performed. When more measurements are made than are needed, the method is said to be overdetermined. This is the case with the error adapter. In robust implementations, the additional measurement is put to use to calculate the same term twice, assuring a valid calculation. The number of calibration measurements required depends on the architecture of the VNA (either three-sampler or four-sampler), as well as the calibration method.

Both the signal flow graph and error adapter assume the impedance presented by port 1 remains unchanged, regardless of the position of the switches shown in Figure 1.3. By incorporating the effects of the switches, the error adapter avoids separate forward and reflected diagrams. This makes the error adapter useful for four-sampler VNAs but invalid for three-sampler ones.

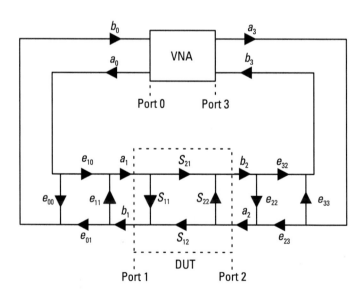

Figure 2.6 The error adapter, made up of eight error terms e_{mn}. Also shown are the forward a_i and reflected b_j signals.

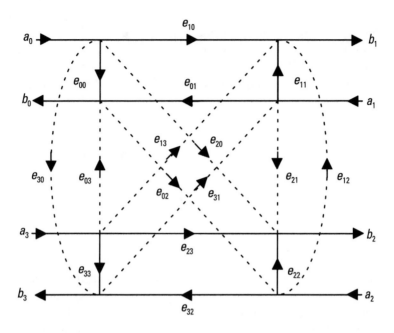

Figure 2.7 Dotted lines show the eight leakage paths in the error adapter of Figure 2.6.

For reference, Table 2.1 shows how to convert between the error adapter and the signal flow graph. For example, e_{11} represents both the forward source match E_{SF} and the reverse load match E_{LR}.

2.4 Calibration Standards

As previously discussed, a valid calibration involves measuring a series of physical elements called standards. Each standard is precisely known as to its magnitude and phase over frequency. By measuring standards in place of the DUT,

Table 2.1
Equating the Terms in the Error Adapter to Those in the Signal Flow Graph

$e_{10} \bullet e_{01} = E_{RF}$	$e_{23} \bullet e_{33} = E_{RR}$
$e_{00} = E_{DF}$	$e_{33} = E_{DR}$
$e_{11} = E_{SF}, E_{LR}$	$e_{22} = E_{SR}, E_{LF}$
$e_{10} \bullet e_{32} = E_{TF}$	$e_{01} \bullet e_{23} = E_{TR}$

the RF path can be calibrated, de-embedding the errors in the test setup. Most calibration techniques use a combination of standards such as a short circuit, an open circuit, a load, and a thru line. Combined, these uncover the test set-up's error. Electronic standards as well as physical standards can be used [3].

2.4.1   Calibration Coefficients

A true open or short reflects 100% of the incident signal. A perfect termination absorbs all the incident energy. A thru connects the two ports together, ideally with no physical length and no loss. In reality, no standard looks ideal. One reason is that the skin depth of the conductor becomes thinner as the frequency increases, adding to conductor loss at higher frequencies.

Correction factors known as calibration coefficients make up for the standards' deficiencies. When entered into the VNA, the calibration coefficients alert the VNA to the deficiencies so that it can correct for them, making the standards appear as perfect. It is worth noting that the most common mistake during the calibration process is improperly entering the calibration coefficients into the VNA. Entry of the coefficients into the VNA should be done with meticulous care and planning.

The value of the calibration coefficient depends on the construction of the probe, as well as the standards. For example, a large pitch between the signal and ground probes will have different calibration coefficients than probes with narrow spacing. The coefficients are particular to a combination of test system probe and standard. If either the coplanar probe or the standard is changed, then so must the coefficient values be. A common mistake is to swap coplanar probe configurations, such as balanced (ground-signal-ground) to unbalanced (ground-signal or signal-ground) without changing the calibration coefficients. Also, placement of the probes on the standards is important. If a coplanar probe does not contact the standard at precisely the right spot, then the calibration coefficients will be slightly off. Probe placement is covered in more detail in Section 3.5.3 in Chapter 3.

Calibration standards are either direct or self-calibrating [4]. The RF behavior of direct calibration standards is well known and usually characterized at the factory. The calibration sheet attached by the manufacturer confirms their performance. Direct standards are essential to uncovering individual error terms in the signal flow graph (see Section 2.3.1). On the other hand, the RF behavior of self-calibrating standards is only partly known. This makes them easier to build either at the factory or in the lab. Self-calibrating standards are well suited to the error adapter where individual error paths are not necessarily sought.

2.4.2 Short

The ideal short totally reflects the incident signal so that its reflection coefficient $\Gamma=1$. After impinging on the short, the incident signal changes phase 180° and travels back toward the source.

All short standards exhibit some inductance, becoming especially significant above 4 GHz. On the Smith chart, this inductance causes the trace to sweep clockwise along the upper-half plane (see Figure 2.8). Note that a short occurs a quarter-wavelength before or after an open.

With coplanar probes, a conducting strip is used to short the signal and ground probes together (see Figure 2.9). In a test fixture, the short standard is a microstrip line whose end is shorted with a via to ground. The via introduces inductance. For this reason, the calibration coefficient inputted into the VNA is inductive. At high frequencies, skin effect contributes some resistance in addition to the inductance.

In the VNA, the inductance of the short can be modeled as a third-order polynomial

$$L(f) = L_0 + L_1 \times f + L_2 \times f^2 + L_3 \times f^3 \tag{2.5}$$

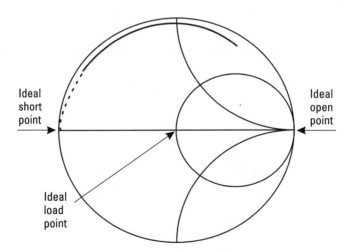

Figure 2.8 Dotted line shows how the short sweeps the edge of the Smith chart due to inductance. When the short has no inductance, the trace never leaves the short point, regardless of frequency.

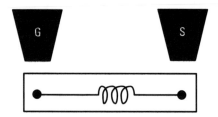

Figure 2.9 Electrical model of a coplanar probe short. The ground and signal probe tips are shorted together on a conducting bar.

where $L(f)$ is in units of H•Hz, L_0 in units of nH, L_1 in 10^{-24}H/Hz, L_2 in 10^{-33}H/Hz^2, and L_3 in 10^{-42}H/Hz^3. At higher frequencies, the distributed nature of the inductance makes the latter terms in (2.5) important.

2.4.3 Open

The open makes no connection between the signal and ground, ideally reflecting the incident signal with no loss. An ideal open would yield a reflection coefficient Γ of unity with 0° of phase shift. In reality, electric fields travel from the signal probe to the ground probe, forming parasitic capacitance. To illustrate, Figure 2.10 shows a coplanar probe open.

On the Smith chart, the open is found at the far right side of the real (or horizontal) axis (see Figure 2.8). As the frequency sweeps, the trace travels clockwise through the capacitive (or lower) half. At a quarter-wavelength, it arrives at the short. The sweep then continues through the inductive (or upper) plane. As an open tends to radiate, this, along with fringing fields, contributes to its loss. Due to loss, the trace spirals inward with each revolution on the Smith chart.

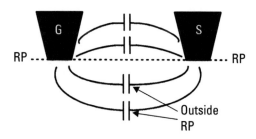

Figure 2.10 A coplanar probe open showing the capacitive electric fields.

Because the capacitance of the open is usually larger than the parasitics of the other standards (short, load, and thru), it is important that the open be accurately characterized. The capacitance found inside the *reference plane* (RP) is de-embedded upon calibration. The capacitance found outside RP is the fringing capacitance. Physically removing the fringing capacitance of an open is difficult. Mathematically neutralizing it using calibration coefficients in the VNA is simpler. A third-order polynomial describes the fringing capacitance

$$C(f) = C_0 + C_1 \times f + C_2 \times f^2 + C_3 \times f^3 \qquad (2.6)$$

where $C(f)$ is in units of F(Hz), C_0 in units of fF, C_1 in 10^{-27}F/Hz, C_2 in 10^{-36}F/Hz^2, C_3 in 10^{-45}F/Hz^3. As with (2.5), its polynomial nature accounts for dispersion. High-impedance DUTs are reflective, so their measurement relies on accurate higher-order terms in (2.6). For low-impedance DUTs, the first-order term C_0 usually suffices.

Characterizing an open standard to find its calibration coefficients is straightforward. First, calibrate the RF test system with a calibration method that does not require a known reflection coefficient (see Section 2.6.3). After calibration, measure the phase shift of the open standard, looking at the VNA's phase display to get its amount of phase shift $\Delta\theta$. To the first order, the effective capacitance C_{eff} of the open should fit the equation

$$C_{\text{eff}} = \frac{\tan\left(\dfrac{\Delta\theta}{2}\right)}{2\pi f Z_0} = \frac{1}{2\pi f X} \qquad (2.7)$$

where

$\Delta\theta$ = the measured phase shift;

f = measurement frequency;

Z_0 = characteristic impedance;

X = measured reactance.

Calibration coefficients cannot correct for coupling or radiation loss, so shielding around the cables or even the probe station helps desensitize the capacitance measurement.

2.4.4 Load

In simplest terms, the load is a terminating resistor mounted between signal and ground probes (see Figure 2.11). When swept over frequency, the load trace ideally swirls about the center of the Smith chart (see Figure 2.8). In reality, parasitic inductance arises due to the interconnecting pads at either end of the resistor. A minor amount of parasitic capacitance also exists between the two pads. Loads greater than 100Ω tend to have larger amounts of shunt capacitance.

The load's impedance is usually designed to match the test system's characteristic impedance Z_0 (usually 50Ω). Any difference between the load and Z_0 will reflect a small part of the incident signal from the load back toward the test system. This reflection adds to any leakage between the reference and test channels inside the VNA. Matching the load to Z_0 lowers the test port's mismatch and prevents such systematic errors from appearing in the directivity measurement.

Of the standards described in this section, the load is the most difficult to build [5]. For balanced coplanar probes, two parallel $100\text{-}\Omega$ resistors form a $50\text{-}\Omega$ load (see Figure 2.12). The resistors are precisely trimmed at the factory using a dc ohmmeter. However, dc trimming does not necessarily improve its high-frequency behavior. Above 26.5 GHz, skin effect increases the return loss and shifts the phase linearly over frequency. For wide-bandwidth calibrations, it is best to use several loads, each tuned for a particular frequency band. When the probe's construction allows, mounting the $100\text{-}\Omega$ resistors at right angles to one another minimizes the coupling between the two.

Another way to think of the load is as an infinitely long transmission line. Applying this idea at millimeter-wave frequencies, a long spiraling line

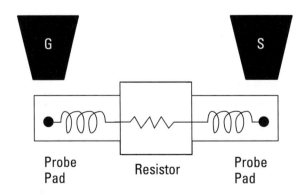

Figure 2.11 Coplanar probing the load. Also shown is its equivalent circuit.

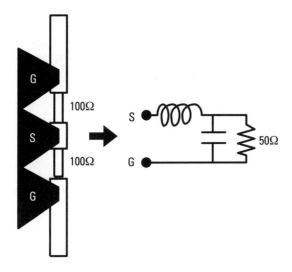

Figure 2.12 A coplanar 50-Ω load. Two 100-Ω resistors in parallel equate to 50Ω. Its equivalent circuit is also shown.

can be used to emulate a load (see Figure 2.13). On the calibration substrate, surrounding the spiraling line with resist attenuates the RF signal and prevents edge coupling.

Do not apply bias when using the load standard; otherwise, it will dissipate power and generate heat. Because the resist material has a thermal coefficient, the load's value changes as it heats. Be sure to calibrate at the DUT's operating temperature. Designing the calibration standards to the test environment will always yield better results (see Section 7.3).

In coaxial calibration kits, the load can be either a fixed 50-Ω standard or a sliding load. The coaxial sliding load can be adjusted to resemble a seamless interface from the coaxial line to the load. Using the sliding load, the coaxial connector is viewed simply as a transmission line mismatch that can be easily calibrated out. Adjusting it to multiple slide positions determines the test system's directivity.

2.4.5 Thru

A thru connects the two RF test ports together. The typical thru realization is a length of 50-Ω line (see Figure 2.14). A good thru line presents the minimum mismatch to the test system's ports, having a constant impedance Z_0 over its length. A zero-length thru is ideal, directly connecting the two probes together. In practice, this is not feasible, especially since test system probes

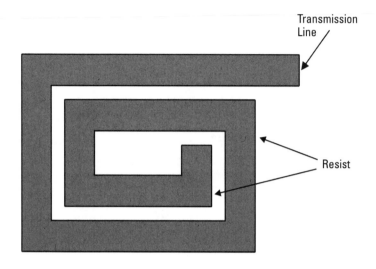

Figure 2.13 At millimeter-wave frequencies, a long 50-Ω transmission line can be used as a load.

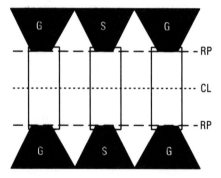

Figure 2.14 A coplanar probe thru, with dotted lines showing the probe's RPs and the *center line* (CL) of the thru.

can easily collide when brought too close together. The result is that the thru's length contributes a parasitic to the calibration, not unlike the capacitance of an open or the inductance of a short. The thru's length should be de-embedded in both magnitude (offset loss) and phase (offset delay) (see Section 2.5).

The design of the thru must be carefully considered, especially a right-angle thru. Problems can arise when the DUT's ports are at right angles to one another. Right-angle transmission lines are capacitive and dispersive in nature. Properly mitering the corner minimizes the capacitance but does not eliminate it. Not only does a right-angle thru not present a pure 50-Ω line impedance, it is also longer. Long thru lines have significant shunt capacitance, degrading the accuracy of calibration methods such as *thru-reflect-line* (TRL) that rely on a well-behaved thru (see Section 2.6.3).

When an IC designer decides the location of the input and output ports of the DUT, he or she unknowingly sets the design of the thru at the same time. Unfortunately, it is the test engineer who must deal with the DUT's port placement.

2.5 Improving the Standards

This section describes two methods of shifting the electrical reference plane beyond the calibration plane. Offset delay shifts the phase of the signal while offset loss shifts its magnitude. Together, the magnitude and phase constitute the signal's complete vector information.

2.5.1 Offset Delay

Offset delay is useful in compensating for differences in length between the physical reference plane and the electrical reference plane. Sometimes the open, short, or load standards are not attached directly at the probe tips. Instead, probe pads and interconnecting lines lead up to the standard. For instance, the load in Figure 2.11 has contact pads at either end that have electrical length as well as physical length. Applying offset delay is a simple way to shift the electrical phase to compensate for the pads.

Offset delay rotates the phase of the signal found at the calibration plane. The phase rotation serves as an electrical delay that can be either positive or negative. Adding (or positive) delay moves the electrical reference plane beyond the calibration plane, while shifting the plane inwards is negative delay. Looking at a phase-only plot on the VNA's display helps to adjust. Offset delay assumes a linear phase and a constant impedance along the line. Extending the phase linearly along a transmission line gives misleading results unless the line is straight and free of discontinuities. Applying offset delay does not affect the signal's attenuation; rather, it remains the same as that at the calibration plane.

Offset delay is based on the one-way travel time of the signal from the calibration plane to the standard. Defining the effective phase velocity v_{eff} as

$$v_{\text{eff}} = \frac{c}{\sqrt{\varepsilon_r}} \qquad (2.8)$$

The offset delay τ_D (in seconds) is

$$\tau_D = \frac{l\sqrt{\varepsilon_r}}{c} \qquad (2.9)$$

where

l = physical length of offset in meters

ε_r = relative permittivity

c = speed of light (m/s)

When the effective phase velocity v_{eff} is known, the electrical length can be converted into physical length by simply multiplying v_{eff} and τ_D.

Use offset delay either through an offset phase ($\Delta\theta$) or a time delay (τ_D). By thinking of a parasitic as a lossless phase length $\Delta\theta$, offset delay can correct for an open's capacitance or a short's inductance. Viewing the parasitic as an offset length Δl, the electrical reference plane can be extended beyond the physical reference plane by an offset phase amount $\Delta\theta$

$$\Delta\theta = \frac{2\pi f \Delta l}{c} \qquad (2.10)$$

which adds phase but does not add loss. Since it assumes that the phase shift $\Delta\theta$ is linear with frequency, this is a good first-order approximation.

Offsetting the electrical reference plane is also possible by adding (or subtracting) a length of the transmission line. The line's length can be defined in terms of its time delay τ_D rather than its physical length l. To avoid loss, the offset line should be short and exhibit a high characteristic impedance Z_0 (500Ω or greater). Establish an offset line by introducing a time delay T_D

$$T_D = \frac{L}{Z_0} \qquad (2.11)$$

where L is the parasitic inductance of the standard (a short or a load). This method is adequate up to 6 GHz.

2.5.2 Offset Loss

To accurately shift the electrical plane along a physical line requires adjustments for loss, as well as phase. Like offset delay, offset loss adds or subtracts the loss of a fictitious length of line.

The most common use of offset loss is to shift the calibration plane from the middle of a thru back toward the coplanar probe tips (see Figure 2.14). The offset loss of a fictitious length of transmission line at 1 GHz is calculated by

$$\text{Offset loss}\left(\frac{G\Omega}{S}\right)_{1\text{ GHz}} = \frac{\text{dB}_{\text{loss}}|_{1\text{ GHz}}\, c\sqrt{\varepsilon_r}\, Z_0}{10\log_{10}(e)l} \tag{2.12}$$

which is in units of $(\text{dB}_{\text{loss}}|_{1\text{ GHz}})$, Z_0 is the characteristic impedance of a line of length l, c is the speed of light, ε_r the relative dielectric constant, and the exponential constant is e.

To calculate the offset loss in terms of a time delay from a fictitious line

$$\text{Offset loss}\left(\frac{G\Omega}{s}\right)_{1\text{GHz}} = \frac{2303\, Z_0\, \text{dB}_{\text{loss}}}{T_D\sqrt{f}} \tag{2.13}$$

where T_D is the delay from the center of the thru to the probe tips, and dB_{loss} is its insertion loss at frequency f. Either (2.12) or (2.13) can be used to electrically shift the attenuation from the calibration plane.

2.6 Calibration Methods

Generally speaking, calibration transfers the quality of the calibration method to the quality of the RF test system. In other words, any error in the calibrated measurement will be due to the calibration method or standards, not to the test system's imperfections. A variety of calibration methods are available to the test engineer. This section explains the limitations of each and briefly covers their theory. Table 2.2 summarizes some of the more popular calibration methods [6]. To best compare, the reader should ask a few questions. Is a complete and thorough knowledge of the calibration standards crucial? How does knowing them any less affect the measurement validity? How repeatable is the method? And lastly, does the method need to consider isolation?

Each calibration technique in Table 2.2 uses a different combination of standards. While the principles of each method apply to virtually any RF media (such as coaxial, coplanar waveguide, and microstrip), examples in this section are illustrated in terms of coplanar probing. All assume a single mode of propagation, the dominant mode. Some calibration methods have redundant information. That is, the same information can be found from more than one measurement. This can be a source of confusion if the same answer is not arrived at with different measurements. In general, there is a tradeoff between the number of standards and accuracy, with more standards leading to better accuracy.

In the literature, three-sampler calibration techniques are denoted with a * (e.g., SOLR*, TRL*, LRM*, and LRRM*). *Short-open-load-thru* (SOLT) is the only method identical to both three- and four-sampler VNAs (see Section 1.2.1). It is possible to use a three-sampler VNA with SOLT by driving one RF source alternately between ports.

Calibration methods can be separated into two types. The simplest type is a response calibration (see Figure 2.15). It corrects solely for scalar errors. The calibration standard used is a short or open element (for one-port or reflection measurements) or a thru element (for two-port or transmission measurements). Simply measure the magnitud with the element inserted and compute the DUT measurement (see Figure 8.17 in Chapter 8). The more accurate type is a vector calibration, correcting for both magnitude and phase. A vector calibration corrects for RF behavior such as mismatches at each port, frequency-response tracking within the VNA, and directivity errors in the system (see Section 1.2 in Chapter 1). A one-port vector

Table 2.2
List of Calibration Methods Available to Three-Sampler (*) and Four-Sampler VNAs

Calibration	Ideal Standards	Nonideal Standards
SOLT	Short, Open, Load, Thru	None
SOLR, SOLR*	Short, Open, Load	Reciprocal Thru
TRL, TRL*	Line	Thru, Reflect
LRM, LRM*	Line, Match	Reflect (Open or Short)
LRRM, LRRM*	Line, Match's Resistance	Reflect, Match's Inductance

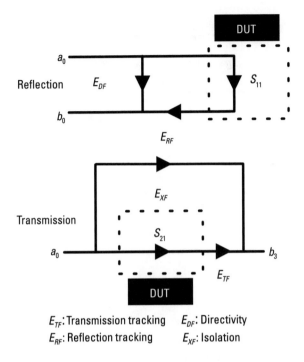

E_{TF}: Transmission tracking E_{DF}: Directivity
E_{RF}: Reflection tracking E_{XF}: Isolation

Figure 2.15 Response calibration for either a one-port (reflection) or two-port (transmission) test system. To find the error terms, the DUT is replaced with a calibration standard.

calibration is performed by mounting a series of calibration standards such as a short, an open, and a load in place of the DUT (see Figure 2.16). Adding a thru to the standards enables a two-port vector calibration.

2.6.1 SOLT

In the early days, precision coaxial standards (shorts, opens, loads, and thrus) were easy to build. This coaxial foundation led to SOLT becoming the most popular calibration method employed today. SOLT relies on well-known standards, all defined along the same reference plane (see Figure 2.17). With SOLT, there is a direct connection between knowledge of the standards' precise RF characteristics and the accuracy of the calibration. Well-known standards bring forth a better SOLT calibration.

A two-port SOLT calibration requires finding the 12 error terms in Figures 2.4 to 2.5. *Short-open-load* (SOL) measurements at each port suffice

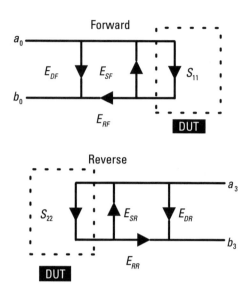

Forward

E_{RF}: Reflection tracking
E_{SF}: Source match
E_{DF}: Directivity

Reverse

E_{RR}: Reflection tracking
E_{SR}: Source match
E_{DR}: Directivity

Figure 2.16 One-port vector calibration, shown for port 1 (forward) and port 2 (reverse).

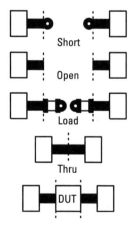

Figure 2.17 Microstrip SOLT standards. The dotted lines denote the calibrated reference planes.

for six terms. A forward reflection calibration finds E_{DF}, E_{SF}, E_{RF} while a reverse reflection calibration E_{DR}, E_{SR}, E_{RR} (see Figure 2.16). An additional thru standard gauges the remaining terms such as the transmission frequency response in each direction, E_{TF} and E_{TR}. With the thru connected, use one port to measure the match presented by the other port. Measuring in each direction provides the load terms, E_{LF} and E_{LR}. After removing the thru and terminating both ports, measuring the insertion loss will reveal any leakage between ports, thereby giving the isolation terms E_{XT} and E_{XR}. This brings the total error terms to 12.

In truth, the 12 calibration measurements give more information than what is needed to determine the 12 terms in the error model. Because three of the error terms in the model are solved for twice, the equation set is said to be overdetermined. This can be a drawback to the technique. Calibration validity comes into question when solving different equations for the same terms yields unequal results.

Although popular, the SOLT method has disadvantages, the principle one being accuracy of the standards. The better the standards are known, the better the calibration. All the standards used in SOLT are direct standards (see Section 2.4.1). Even small deviations from ideal can lead to large errors, manifested in regions of the Smith chart far from the calibration standard [7]. Furthermore, accurately characterizing the SOLT standards becomes laborious at frequencies above 20 GHz [8]. The distributed nature of the standards complicates generating high-frequency models of the SOL. As mentioned, the ideal thru connects the two ports perfectly. Because this is rarely possible, applying offset delay and offset loss are the best ways to improve the thru.

2.6.2 SOLR

The *short-open-load-reciprocal* (SOLR) thru method closely parallels the SOLT [9, 10]. SOLR is preferred when the system requires the use of an imperfect thru, such as a right-angle thru (see Figure 2.18). By taking advantage of redundant information in the SOL measurements, only the thru's offset delay need be known ahead of time. For SOLR, the thru must be reciprocal, that is, symmetrical so that $S_{12}=S_{21}$, assuming reciprocity is what differentiates SOLR from other techniques.

Note that no matter how well mitered, the corner of a 90° thru can lead to reflections. For instance, in *coplanar waveguide* (CPW) corner reflections arise from slot-line modes launched at the bend [11]. Parallel-plate modes and surface waves can also appear in CPW bends to a lesser degree.

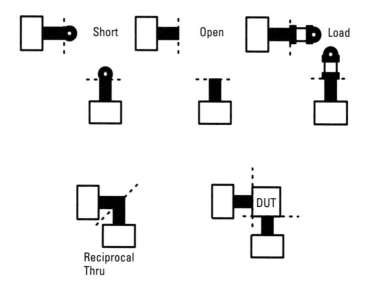

Figure 2.18 Microstrip SOLR standards. The dotted lines denote the reference planes.

2.6.3   TRL

Transmission lines are simple to understand and easy to fabricate. Their physical dimensions and the board material decide their characteristic impedance. Because it is based on a transmission line standard, TRL is a powerful method [12]. Figure 2.19 shows TRL standards in microstrip. A short transmission line serves as the thru. When the offset delay is set to zero, the thru's midpoint sets the electrical reference plane. In general, the line lengths leading to the DUT should be the same length as the thru. Known to within +/–90° is the thru's phase length θ. The reflect standard can be either an open or a short. Only the sign of its reflection coefficient need be known.

The reflect does not have to be a perfect open or short, although the best results have a $|\Gamma|_{\mathrm{REFLECT}}$ close to 1. The reflect's phase should be generally known to within +/–90°. When selecting a reflect, keep in mind an open operates over a broader bandwidth than a short. Be sure to use the same reflect standard for both ports. When its phase is well known, the reflect can be used to define the electrical reference plane instead of the thru.

The one precision standard is the line. Its characteristic impedance Z_0 sets the reference impedance for the entire RF test system. For good RF measurements, the line's impedance should be precisely defined; otherwise, normalization is required. For instance, say the characteristic impedance of the line is not 50Ω but instead 48Ω. Because the measurements are referenced to

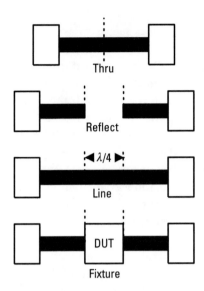

Figure 2.19 Microstrip TRL standards. The reference plane is defined at the center of the thru (dotted lines).

the characteristic impedance of the line, the measured S-parameters must be normalized from 48Ω to 50Ω. Mathematically, normalizing the data is done using a 50:48 transformer at the input port and a 48:50 transformer at the output port. This is a simplistic method of normalization since it will not account for reflections at the 48-Ω to 50-Ω interface.

TRL sets the test system's Z_0 to be equal to the impedance of the line standard. However, tolerances in the conductor etch fabrication can result in the line's impedance being slightly different than Z_0. Since the DUT measurements are defined in terms of Z_0, error will result. The TRL's inability to compensate for imperfect transmission lines is its principal weakness.

Line loss and dispersion will impact the impedance of the line's Z_0. The loss in the line standard is partly due to the metal's conductivity, a function of the skin depth. Making assumptions about transmission line symmetry can sometimes compensate for calibration errors due to line loss [13].

The line standard is usually good over an 8:1 (frequency span:start frequency) frequency range. Wider bandwidths require more than one line standard. Overlap the bandwidths of each line to ensure calibration accuracy at the band's edges [14]. At low frequencies, the line standard becomes prohibitively long and lossy. If the test ports are hard-mounted (i.e., fixed), then the varying thru and line lengths will pose a problem. Meandering transmission lines are sometimes suggested as a work-around for the longer line; however, meandering exhibits excess capacitance that degrades Z_0.

In order to present different phases during calibration, the thru and line standards *must* be different lengths. The thru's optimum length is 90°, or $\lambda/4$ of the center frequency. In general, the line's length should differ from the thru's by 20° to 160°. Note this is in $\lambda/2$ (or 180°) increments. That is, the line's length could just as well be 200° to 340° longer than 20° to 160° longer. Neither the thru nor the line should be a half-wavelength ($\lambda/2$) increment of the other since a signal 180° out-of-phase from another yields the same phase information. A line that is $\lambda/2$ longer than a thru will electrically behave as a zero-length line standard, or an open. TRL computation error will increase as the thru's length approaches $\lambda/2$ of the measurement frequency [15, 16].

After a successful TRL calibration, the reference plane is not necessarily in-phase with the RF signal coming out of the test port. Due to an arbitrary root choice in the TRL computation, the calibrated reference plane will either be in-phase (0° offset) or 180° offset from the RF test signal [17].

TRL has 12 unknowns in its signal flow graph, found with 16 measurements. Instead of using a signal flow graph, the number of measurements can be reduced by using an error adapter. The problem then reduces to eight error terms, four on each side of the DUT (see Figure 2.6). Ten measurements (S_{11}, S_{22}, S_{12}, S_{21} for both the thru and line, S_{11} and S_{22} for the reflect) obtain the eight unknown error adapter terms. Because 10 measurements are used to find eight unknowns, the two extra measurements can reveal other information such as the reflection coefficient of the reflect $|\Gamma|_{REFLECT}$ or the propagation constant β of the line. Knowing β allows the reference plane to be accurately shifted in both magnitude and phase, solving the root problem.

A number of variations on the TRL technique exist. With the *line-reflect-line* (LRL) method, the test system's impedance is set by the characteristic impedance of two well-defined lines rather than a single thru [18]. One of the LRL lines is about $\lambda/4$ longer than the other. The advantage of LRL is that the input and output connectors need not be the same, avoiding adapter removal. *Thru-reflect-attenuate* (TRA) replaces the line standard with a wide-band attenuator, avoiding TRL's $\lambda/2$ resonance problem [19]. In *thru-short-delay* (TSD), either the short or the delay is the well-characterized standard [20].

TRL works best with a four-sampler VNA. By simplifying the error terms affected by switching, it can also be used with a three-sampler VNA (see Figure 1.3). However, this method, called TRL*, sacrifices some accuracy [21]. In noncoaxial media such as CPW, TRL is the preferred method due to the self-calibrating nature of the standards [22].

2.6.4 LRM

Similar in form to TRL is *line-reflect-match* (LRM) [23]. The line and reflect are analogous to the thru and reflect standards in TRL, the difference being that LRM uses a precision match (or load) to define the system's characteristic impedance Z_0. Again, either an open or a short can serve as the reflect (see Figure 2.20). As in SOLT, the load must be well defined; otherwise, the calibration sensitivity is degraded. The line standard sets the electrical reference plane.

Contrary to SOLT, neither the open's capacitance nor the short's inductance need be known before calibration due to the math involved [24]. Using the same reflect and load standards to calibrate both ports gives the most accurate results. Upon completion of an LRM calibration, the reference plane for the two ports is at the center of the line.

The LRM's unique aspect, the match, is also its weakness. As discussed in Section 2.4.3, parasitic inductance corrupts the load. The inductance of the interconnecting pads at either end of the match results in LRM calibration error. When this inductance is accounted for, the accuracy of LRM is comparable to TRL. To minimize inductance, ensure the conductor traces leading to the match are short in length.

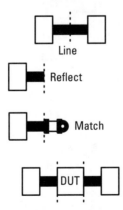

Line

Reflect

Match

DUT

Figure 2.20 Microstrip LRM standards. The dotted line denotes the reference plane. Either an open or a short can serve as the reflect. For best results, use the same reflect and match standards for both ports.

When a precise, well-defined load is available, the LRM method has a high degree of accuracy. Otherwise, when only a short, well-known thru can be constructed, TRL has the advantage. To trade off their strengths, SOLT can be performed at lower frequencies and TRL at higher frequencies. LRM functions as a combination of the two. It generally has better accuracy than the SOLT method, even when using the same standards to perform both methods.

2.6.5 LRRM

Line-reflect-reflect-match (LRRM) is nearly identical to LRM but with one advantage. It uses the additional reflect to alleviate LRM's principal flaw, inductance in the match (see Figure 2.21). An added benefit is that the match does not have to be in the system's characteristic impedance Z_0.

Like LRM, LRRM works with fixed-probe separation and uses the same set of standards as SOLT. Unlike SOLT, the short and open in LRRM do not have to be well defined. As with TRL, characterizing the line is limited to finding its electrical delay and loss. All other information about the standards is not critical to getting a quality calibration.

Accurately characterizing the inductance of the match standard is the primary benefit offered by LRRM, overcoming a weakness in LRM. RF

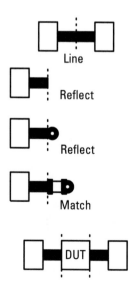

Figure 2.21 LRRM microstrip standards.

measurement of the match is performed on only one port, so no discrepancies arise from using different match standards on opposite ports. LRRM also avoids another problem. When the same standards are used to calibrate both ports, probe placement error arises when moving the same standard from port to port, varying the standard's parasitic inductance amount. Calibrating with one match standard measured on one port eliminates differences in inductance measured on two ports.

LRRM finds the match's inductance by focusing on the open, since it is the most repeatable standard. Following calibration, the open shows no conductance. If, after calibration, the measurement displays negative conductance, then the match has some inductance. LRRM iteratively adjusts the match's calibration coefficient until the open's conductance equals zero.

As with any of these methods, there are disadvantages. LRRM applies assumptions about the match. Namely, the match measured on one port is assumed to behave identically on the other port. This can lead to error in the return loss measurement (S_{11} and S_{22}) when the two ports are not actually the same.

Other methods comparable to LRRM are available [25]. Rather than a second reflect, a similar method employs a second line to correct for match imperfections and losses in the line [26].

2.7 Verification

Following calibration, it is important to ensure the calibration has been performed and computed properly. This is the purpose of verification, to confirm the calibration and bring to light any residual errors in the test system that escaped the calibration process. Performing a verification after every calibration is generally good practice.

Residual errors in the VNA can originate from source or load mismatches, the transmission tracking, or the reflection tracking (that is, the frequency response of the forward and reverse signal paths), or the directivity of components used in the test system. A few simple tests can reveal residual errors in a coaxial test system [27]. Remeasuring the short and open standards, this time mounted at the end of a length of line, is one way. First measure a short at the end of a long line. Viewed on a fine-enough scale, ripples will appear in the return loss measurement. Place this trace in the VNA's memory and then measure the return loss of the open standard at the end of the same length of line. By overlaying the two traces, the envelope of ripple gives the amount of residual error in the test system. Conduct a second test by connecting a load at the end of a transmission line of characteristic impedance Z_0. The transmission line between the load and test port should be long enough to generate ripples in the return loss. Follow this measurement by connecting an identical load at the test port with no transmission line length in between. It should yield smaller ripples than those measured with the load at the end of a long length of line.

One form of verification is to compare the calibration results to the results from a previously made, known-to-be-good calibration. This benchmarks the just-completed calibration to a known good one, gauging the test system's drift over time. Checking a calibration using verification elements is more common than comparing calibration results.

2.7.1   Verification Elements

In theory, any RF element can be used to verify a calibration. Anything accurately characterized a priori is a candidate. Verification elements are not the standards used in the calibration process, since remeasuring the calibration standards again simply proves their repeatability. Remeasuring them after calibrating should yield the standards' calibration coefficients initially entered into the VNA. This is of particular interest to SOLT, which relies on the quality of the standards. In any calibration method, remeasuring the open after calibration should show a small capacitance. This capacitance value should equate to the open's calibration coefficient. Note that neither the short nor the open should at any time travel outside the Smith chart. With TRL or LRM, remeasuring the open can highlight calibration problems, especially when the open incorrectly exhibits gain. When remeasuring the thru standard, S_{12} should equal S_{21}.

To verify a scalar calibration, try remeasuring a wideband attenuator such as a 10-, 20-, or 30-dB pad. To verify a vector calibration, either an open or a long thru are convenient verification elements. Looking at the reflection coefficient of a long, open-circuited line $|\Gamma|_{OPEN}$, the sweep should begin at low frequencies on the right axis of the Smith chart and spiral clockwise inwards (see Figure 2.8). With a long thru line connected between the test ports, the phase of an insertion loss (transmission) measurement should show linear behavior. Other good verification elements are high-Q capacitors and inductors. Because of their low loss, they should smoothly hug the edge of the Smith chart during a frequency sweep. Similarly measuring a 100-Ω resistor should show $|\Gamma| = 0.5$. One interesting method uses a resonator as a verification element [28]. The propagation constant γ of the resonator is found from an uncalibrated measurement. Then it is compared to the resonator's γ measured on a calibrated system. The difference in γs gives the amount of error that escaped calibration. When remeasuring verification elements, keep in mind that the reference plane may have shifted due to the probe contacting a slightly different point.

As another verification element, consider an open that is different from the one used during calibration, such as an open at the end of a long length of line. Plotting the return loss on a Smith chart, the measurement trace of the longer open should start where the sweep of the shorter open calibration standard stopped. This is because calibration will shift the reference plane by the length of line of the open standard used in the calibration process. Sweeping other verification elements in the same fashion should give the same result.

2.7.2 Verifying with a Reference or "Golden" Unit

The simplest way to verify a calibration is with a "golden" unit. This is a known good DUT regularly used to check the calibration. The golden unit's response, while perhaps not ideal, is measured when the test system is known to be in good working order. Periodically remeasuring with the golden unit serves as a simple test system check. Inserting a reference or golden unit in place of the DUT is often used to qualify a manufacturing test setup. The golden unit is a good DUT that was pulled from the test line.

On a manufacturing line, measurement consistency can be more important than measurement accuracy. In this case, repeatability of the results is most important. On production lines, remeasuring the golden unit over and over eventually wears it out. As it wears down, the characteristics of the golden unit itself will change. Eventually, it is replaced with another golden unit, undoubtedly having different RF characteristics. This makes measurement repeatability and reproducibility difficult to trace over time.

Here are a few other points to consider when using reference units. Qualifying the test system with a golden unit may not necessarily reveal all the problems with a calibration. Also, when using an active device as the golden unit, slight changes in the bias point will alter the unit's RF impedance presented to the test system. This makes bias stability important. Despite the disadvantages, verifying with a reference unit is an adequate check of the test system's stability.

2.8 Isolation

Isolation refers to the amount of RF signal that leaks between paths or components within an RF test system. Deciding when an isolation calibration is required depends on the test system's architecture [29]. Isolation becomes an issue in the most physically confined parts of the test system. Signal leakage internal to the VNA is the best example. Specifically, the source and receiver paths closely located inside the VNA must be well separated in terms of RF. After the VNA, the next most susceptible area is at the test system-to-DUT interface. When the DUT's input and output pads are closely together, the test probes can couple from one to another. Two coplanar probes can couple to one another when in close proximity during RFIC or MMIC testing [30].

Recall the E_X terms that represent isolation error in the signal flow graph shown in Section 2.3.1 (see Figures 2.4 and 2.5). An isolation calibration step adds more terms to the signal flow graph (see Figure 2.7) [31, 32]. An isolation calibration requires at least one additional calibration standard to measure the leakage paths. The simplest way to quantify a VNA's internal leakage is to terminate each test port with a 50-Ω load and measure S_{12} and S_{21}. This reveals the VNA's internal isolation.

It is difficult to correct for leakage occurring due to the presence of the DUT. During testing, the test system and DUT will interact. High reflections from the DUT's ports can induce even higher leakage from the VNA's source to its receiver. In such events, the characteristics of the DUT directly impact the test system's isolation. Even more disconcerting, inserting a similar but not identical DUT can change the value of isolation. Varying isolation results lead to DUT measurement error.

For the ideal isolation calibration, the reflection coefficient of the isolation standard $|\Gamma|_{ISOLATION}$ should be identical to that presented by the DUT $|\Gamma|_{DUT}$. In this way, the effect the DUT has on the system isolation can be properly calibrated out. However, fabricating an isolation standard to match the impedance of the DUT's port may not be practical.

Radiation is not the only cause of RF leakage. In test fixtures, any conductor in or around the DUT can serve as a leakage path. When mounting the DUT on a *printed-circuit board* (PCB), the ground plane of the PCB can qualify as a leakage path. Poor DUT grounding to the PCB induces multiple ground currents in the DUT's ground plane (see Section 8.7.1 in Chapter 8).

There is a fundamental rule when performing isolation measurements. The test system's isolation must be better than the isolation of the DUT being measured. For adequate resolution, the test system's isolation should be at least 3–5 dB better than that of the DUT. The quality of the DUT's isolation often decides whether or not to perform an isolation calibration. For example, a bandpass filter has high out-of-band signal rejection, so the test system must have high out-of-band isolation. On the other hand, high-power amplifiers need some amount of port-to-port isolation to avoid oscillation.

2.9 Traceability

The measurements from a calibrated test system are only as reliable as the calibration itself. For universal acceptance, the calibration standards or the calibration itself should be traceable to quantities agreed on by the industry. In this way, all high-frequency measurements can uniformly relate to one another regardless of where they are made.

Short, open, load, and thru standards are used in a variety of calibration methods. Each method (i.e., SOLT, TRL, and LRM) relies on one or more of the standards being accurate. Electrical modeling of the standards can be time-consuming. Mechanically, they wear out from being continually connected and reconnected. When well-defined standards are unavailable, traceability of the calibration result may be a better option.

Figure 2.22 illustrates the many ways a calibrated VNA test system can be traced back to a standard. Benchmarking either the standards or the calibration outcome ensures measurement accuracy and repeatability [33]. The National Institute of Standards and Technology (NIST) holds references for calibration standards such as for the coaxial SOL. Microcalorimeter systems regularly check these standards. NIST specifies the least amount

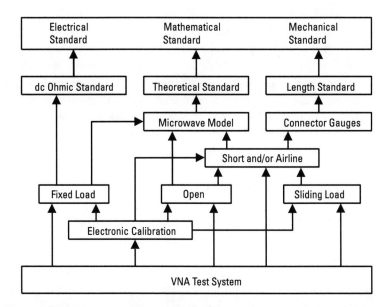

Figure 2.22 The ways to correlate a calibrated VNA test system to universal calibration standards.

of measurement uncertainty possible, usually to within a few tenths of a percent. On-wafer calibration standards, on the other hand, are not defined by NIST. Rather, NIST defines the quality of the on-wafer calibration results.

2.10 Repeatability, Reproducibility, and Accuracy

This section casts light on the distinction between repeatability and reproducibility of RF measurements. Consider what happens when the same RF measurement is performed on the same DUT over and over again, one measurement right after another. A slight difference registers each time. This is *repeatability*. It reveals the test system's systematic errors, as well as its short-term, random errors such as noise. A month later, try remeasuring the same DUT. During the intervening month, the test system may have been recalibrated, system operators may have changed, or perhaps the DUT is being remeasured on another identical test system. The difference gives the *reproducibility*. One source of measurement reproducibility error is due to thermal drift in the test equipment.

Accuracy is how close the measurement is to its true behavior. In general, any well-known element can be used to check the test system's accuracy. For instance, following calibration, remeasuring the calibration standards should yield the calibration coefficients inputted into the VNA.

An operator's probing technique can have an impact on both repeatability and reproducibility. Methods are available to quantify the variability in an operator's measurement technique [34]. Because of its excellent stability, measuring a coplanar probe open is one way. Assuming the probes are well taken care of, its return loss $|\Gamma|_{OPEN}$ measured over time should yield the same result.

2.11 Calibration Tips and Tricks

This section describes typical situations encountered during calibration, illuminating the reader to common pitfalls. It also gives quick pointers to help avoid some of the more routine mistakes.

Are the Calibration Coefficients Correctly Entered?

Improperly inputting the calibration coefficients into the VNA is one of the most common user mistakes. In addition to simply inputting them wrong, the coefficients reflect the degree of accuracy sought. When developing VNA calibration coefficients, four terms, L_0 to L_3, describe the short-circuit inductance. Using only the L_0 term gives less accuracy as the frequency increases. The higher-order terms, L_1 to L_3, account for millimeter-wave parasitic effects.

How Does Calibration Remove Imperfections?

Physically, test system flaws are never actually removed. During test, reflections arising from mismatch between the test ports and the DUT continue to bounce back and forth even after calibration. Physically, the reflections still remain. The act of calibration mathematically de-embeds impedance mismatches in the RF paths. After calibration, any further test system mismatches are attributed to the DUT, even when they result from interaction between the test system and DUT.

One way to suppress mismatch reflections at the ports is to connect an attenuator in-line between the test system and the DUT, calibrating with the attenuator installed. Unfortunately, attenuators reduce the test system's dynamic range. They can also cause problems with calibration methods other than SOLT.

Which SOLT Standard Is the Most Important?

The answer depends on the DUT. For example, small-signal amplifiers tend to have 50-Ω input and output impedances. This makes the load standard the most important. Bandpass filters behave either as a short in their bandpass or an open out of band. Depending on the band, either the short or open is more important.

Where Are Mismatches in the Test System Coming From?

Sharp resonances on a log-magnitude VNA display indicate a fundamental problem in the test system. The electrical length near the DUT is too short to bring about high-Q behavior, hence such problems are not usually close to the DUT. To locate them, compare an open or a short measurement to that of a load. The load should yield a noticeably better return loss. Using this technique, trace the problem through points of the test system.

Why Does the VNA Trace Go Outside of the Smith Chart?

When the trace sweeps outside the Smith chart, it can be for one of two reasons. Incorrect calibration coefficients are often the case. Either the calibration coefficients were improperly entered into the VNA or the wrong ones were recalled from VNA memory. Poor contact to either the DUT or the calibration standards is another strong possibility.

Inputting the wrong short or open calibration coefficient value rotates the Smith chart about the load point. For instance, inputting too low a value for the open's capacitance rotates the Smith chart counterclockwise. When this happens, measuring high-Q inductors will exhibit a low Q and high-Q capacitors will appear to have negative resistance.

Why Do the Open and Short Standards Sweep Along the Outer Edge of the Smith Chart?

When sweeping over frequency, the open and short standards rarely appear as a single point on the Smith chart. Remeasuring the open and short standards after calibration should yield their calibration coefficients. These coefficients were initially entered into the VNA to correct for the imperfections in the standards. A VNA trace of the short will sweep clockwise from the short point across the inductive (upper) half due to its parasitic inductance (see Figure 2.8). Likewise, the capacitance of the open causes it to sweep clockwise from the open point across the capacitive (lower) half.

Another reason for this type of effect may be due to the reference plane definition. Calibration coefficients correct for parasitics as they appear at the reference plane. When coplanar probing, the probe may slide past the intended touchdown point, overshooting it [see Figure 2.23(b)]. In this case, the VNA is expecting the thru to have a 1-ps delay [see Figure 2.23(a)]. Probe overtravel shortens the thru standard. By overshooting the thru, the calibrated reference plane of each port will overlap into the opposite port. The thru delay entered into the VNA's calibration coefficients is then too long. Sweeping the edge of the Smith chart can be the result of the wrong thru delay in the VNA.

Why Is the Measurement Not Repeatable?

For subtle bouncing in the measurement trace, try increasing the sweep averaging. If the measurement exhibits large variations, try contacting the load standard while uncalibrated. The return loss of the load should be at least 6 dB to 10 dB better than the return loss of either the open or the short. Using a combination of coplanar and coaxial standards, apply this rule-of-thumb and trace any faults through the test system.

2.12   Summary

Calibration de-embeds systematic errors from the test system, ranging from cable and probe losses to connector reflections and VNA imperfections. In

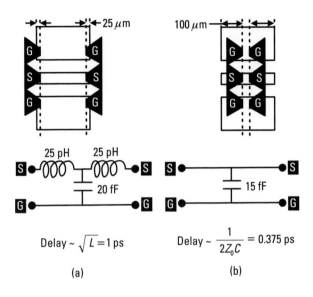

Figure 2.23 (a) Setting the proper coplanar probe position of the thru; and (b) overshooting the thru will shorten the delay.

practice, choosing which calibration method often boils down to which standards can be fabricated reliably. At high frequencies, if good line definition is all that is possible, then choose TRL. When high-quality, high-frequency standards are available, choose a method such as SOLT. LRRM has found wide use for on-wafer, fixed-pitched probes.

Always take care to physically place the probes at the same reference plane, keeping in mind the margins of error in repeatability and reproducibility. In general, the quality of a calibration depends on the quality of the standards. To check the calibration, be sure to employ verification elements.

References

[1] Hewlett-Packard, "Network Analysis: Specifying Calibration Standards for the HP 8510 Network Analyzer," *Hewlett-Packard Product Note 8510-5A*, 1997.

[2] Rehnmark, S., "On the Calibration Process of Automatic Network Analyzer Systems," *IEEE Trans. on Microwave Theory and Techniques*, Vol. 22, No. 4, 1974, pp. 457–458.

[3] Wong, K., and R. Grewal, "Microwave Electronic Calibration: Transferring Standards Lab Accuracy to the Production Floor," *Microwave Journal*, Vol. 37, No. 9, 1994, pp. 94–105.

[4] Eul, H., and B. Schiek, "Reducing the Number of Calibration Standards for Network Analyzer Calibration," *IEEE Trans. on Instrumentation and Measurement*, Vol. 40, No. 4, 1991, pp. 732–735.

[5] Jain, N., and D. Wells, "Design of a DC-to-90-GHz Resistive Load," *IEEE Microwave and Guided Wave Letters*, Vol. 9, No. 2, 1999, pp. 69–70.

[6] Cascade Microtech, "A Guide to Better VNA Calibrations for Probe-Tip Measurements," *Cascade Microtech Technical Brief*, 1994.

[7] Glasser, L., "An Analysis of Microwave De-Embedding Errors," *IEEE Trans. on Microwave Theory and Techniques*, Vol. 26, No. 5, 1978, pp. 379–380.

[8] Okamura, W., M. DuFault, and A. Sharma, "A Comprehensive Millimeter-Wave Calibration Development and Verification Approach," *IEEE Microwave Symposium Digest*, 2000, pp. 1477–1480.

[9] Basu, S., and L. Hayden, "An SOLR Calibration for Accurate Measurement of Orthogonal On-Wafer DUTs," *Cascade Microtech Application Note*, 1995.

[10] Ferrero, A., and U. Pisani, "Two-Port Network Analyzer Calibration Using an Unknown Thru," *IEEE Microwave and Guided Wave Letters*, Vol. 2, No. 12, 1992, pp. 505–507.

[11] Wu, D., et al., "Full-Wave Characterization of the Mode Conversion in a Coplanar Waveguide Right Angled Bend," *IEEE Trans. on Microwave Theory and Techniques*, Vol. 43, No. 11, 1995, pp. 2532–2538.

[12] Engen, G., and C. Hoer, "Thru-Reflect-Line: An Improved Technique for Calibrating the Dual Six-Port Automatic Network Analyzer," *IEEE Trans. on Microwave Theory and Techniques*, Vol. 27, No. 12, 1979, pp. 987–993.

[13] Herman, H., et al., "Millimeter-Wave Deembedding Using the Extended TRL (ETRL) Approach," *IEEE Microwave Symposium Digest*, 1990, pp. 1033–1036.

[14] Marks, R., "A Multiline Method of Network Analyzer Calibration," *IEEE Trans. on Microwave Theory and Techniques*, Vol. 39, No. 7, 1991, pp. 1205–1215.

[15] Heymann, P., H. Prinzler, and F. Schnieder, "De-Embedding of MMIC Transmission-Line Measurements," *IEEE Microwave Symposium Digest*, 1994, pp. 1045–1048.

[16] Kostevc, D., "Simple Extension of TRL Calibration Method of VANA," *Electronics Letters*, Vol. 31, No. 8, 1995, pp. 634–635.

[17] Pantoja, R., et al., "Improved Calibration and Measurement of the Scattering Parameters of Microwave Integrated Circuits," *IEEE Trans. on Microwave Theory and Techniques*, Vol. 37, No. 11, 1989, pp. 1675–1680.

[18] Hoer, C. and G. Engen, "Calibrating a Dual Six-Port or Four-Port for Measuring Two-Ports with Any Connectors," *IEEE Microwave Symposium Digest*, 1986, pp. 665–658.

[19] Soares, R., and P. Gouzien, " Novel Very Wideband 2-Port S-Parameter Calibration Technique," *Microwave and Optical Technology Letters*, Vol. 3, No. 6, 1990, pp. 210–212.

[20] Brubaker, D., and J. Eisenberg, "Measure S-Parameters with the TSD Technique," *Microwaves & RF*, Vol. 24, No. 11, 1985, pp. 97–155.

[21] Metzger, D., "Improving TRL* Calibrations of Vector Network Analyzers," *Microwave Journal*, Vol. 38, No. 5, 1995, pp. 56–64.

[22] Rubin, D., "De-Embedding mm-Wave MICs with TRL," *Microwave Journal*, Vol. 33, No. 6, 1990, pp. 141–146.

[23] Davidson, A., K. Jones, and E. Strid, "LRM and LRRM Calibrations with Automatic Determination of Load Inductance," *Cascade Microtech Application Note*, 1995.

[24] Davidson, A., E. Strid, and K. Jones, "Achieving Greater On-Wafer S-Parameter Accuracy with the LRM Calibration Technique," *Cascade Microtech Application Note*, 1995.

[25] Kostevc, D., and J. Mlakar, "Three-Sampler Network Analyzer Calibrations," *Microwave Journal*, Vol. 42, No. 7, 2000, pp. 88–94.

[26] Williams, D., and R. Marks, "LRM Probe-Tip Calibrations Using Nonideal Standards," *IEEE Transactions on Microwave Theory and Techniques*, Vol. 43, No. 2, 1995, pp. 466–469.

[27] Dunsmore, J., "Techniques Optimize Calibration of PCB Fixtures and Probes," *Microwaves & RF*, Vol. 34, No. 11, 1995, pp. 93–98.

[28] Woodin, C., and M. Goff, "Verification of MMIC On-Wafer Microstrip TRL Calibration," *IEEE Microwave Symposium Digest*, 1990, pp. 1029–1032.

[29] Butler, J., et al., "16-Term Error Model and Calibration Procedure for On Wafer Network Analysis Measurements," *IEEE Microwave Symposium Digest*, 1991, pp. 1125–1127.

[30] Pla, J., W. Struble, and F. Colomb, "On-Wafer Calibration Techniques for Measurement of Microwave Circuits and Devices on Thin Substrates," *IEEE Microwave Symposium Digest*, 1995, pp. 1045–1048.

[31] Silvonen, K., "LMR 16—A Self-Calibration Procedure for a Leaky Network Analyzer," *IEEE Trans. on Microwave Theory and Techniques*, Vol. 45, No. 7, 1997, pp. 1041–1049.

[32] Butler, J., et al., "16-Term Error Model and Calibration Procedure for On-Wafer Network Analysis Measurements," *IEEE Trans. on Microwave Theory and Techniques*, Vol. 39, No. 12, 1991, pp. 2211–2216.

[33] Cascade Microtech, "Technique Verifies LRRM Calibrations for GaAs Measurements," *Cascade Microtech Technical Brief*, 1994.

[34] Raggio, J., "Study Contrasts Different Wafer Probe Test Sites," *Microwaves & RF*, Vol. 30, Vol. 5, 1991, pp. 213–216.

3

Coplanar Probes

CPW probes, also referred to as coplanar probes, are the method of choice for launching RF signals onto and off of a wafer. This chapter explains their construction, RF characteristics, and proper usage.

3.1 Theory of CPW

Before describing coplanar probes, an overview of CPW theory is in order.

There are two types of CPW: *conductor-backed* (CBCPW) and non-conductor-backed (see Figure 3.1). Non-conductor-backed CPW is the same as normal CPW, which has no metal on the bottom or back of the substrate. Although it has better RF behavior than CBCPW, CPW substrates are not as common. Depositing metal on the back provides mechanical strength and thermal dissipation. Electrically, the conductor backing on CBCPW excites three modes of electromagnetic wave propagation: CPW, microstrip, and parallel-plate [1]. Plating the backside of the wafer causes the CPW's signal line to resemble microstrip, so that a microstrip-like mode emerges. Vias usually connect the substrate's backside ground to the CPW grounds on top. However, the top and bottom grounds are at different potentials because of the via hole's inductance, causing a parallel-plate mode to arise.

Energy leakage in the CBCPW is a mixture of the three modes, creating surface waves (see Figure 3.2) [2]. Both parallel-plate and surface waves

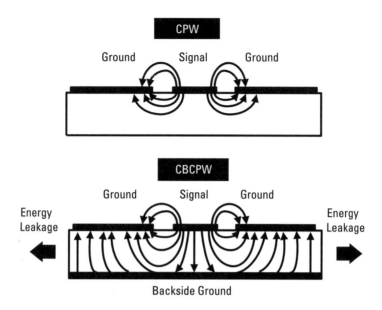

Figure 3.1 Electric fields in CPW and CBCPW. The top figure shows the CPW mode. In CBCPW (bottom figure), the field polarization transforms the CPW mode in the center to microstrip mode at the edges, leaking energy.

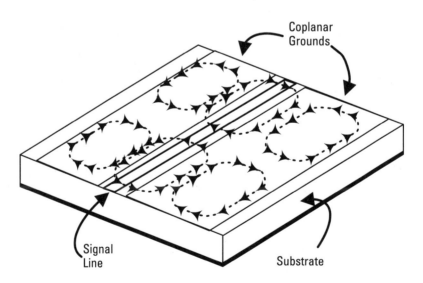

Figure 3.2 Dotted lines showing electric fields behaving as surface waves on a CPW line.

take energy away from the desired CPW mode. One area where energy can be transformed into surface waves is at the coplanar probe-to-wafer transition. This is the point where the probe tips touch the die's pads.

Surface waves arise above a critical frequency and on thicker substrates. To suppress them, connect the coplanar grounds with air bridges across the signal line at judicious points. If the IC process does not allow air bridges, then vias to the backside are another option.

Of the three modes, the desired one is CPW. The mode that dominates depends on the line geometry, the substrate resistivity, and the signal frequency. A wider distance (or slot) between signal and ground lines increases the influence of substrate-induced modes. Conductive, lossy substrates (such as highly doped Si substrates) will behave differently than semi-insulating gallium arsenide (GaAs) ones (see Section 6.1 in Chapter 6). Generally, the CPW mode dominates when the spacing between the CPW grounds is less than $\lambda/10$ [3].

Keep in mind that moding will affect calibration. The calibration equations are premised on a single mode propagating, the dominant one. When energy couples to other modes, calibration will likely fail. Calibration methods such as TRL are especially vulnerable [4].

Regardless of whether the backside is metalized or not, the current density on the CPW signal line is not uniform across its surface. Here, current crowds to the edges of the signal line, even more so than in a microstrip line of comparable width. The edges of the line are rougher than the conductor's surface. Compared to a similar microstrip line, a CPW line with rough edges will have greater insertion loss. CPW's advantage over microstrip is that the CPW line width is independent of the line's impedance, made possible by adjusting the CPW's slot width. A wider line provides more area, lowering the conductor's high-frequency loss.

3.2 Mechanical Construction

Figure 3.3 shows the typical construction of a coplanar probe. The tips are made of flexible beryllium-copper (BeCu) to keep them from digging into the die's probe pads [5]. BeCu is optimal for probing gold pads on fragile GaAs wafers. Not only is it flexible, it also offers low contact resistance. On the other hand, tungsten (W) tips are firmer, breaking through oxide film on aluminum pads to make good electrical contact. During tests, aluminum oxide buildup can vary the contact resistance. While excessive overtravel of the probes can break through the oxide, it causes the probes to wear out

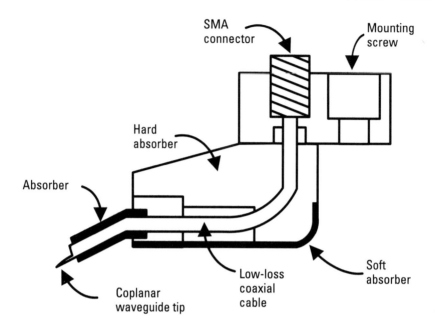

Figure 3.3 Coplanar probe construction.

sooner. Furthermore, the pads must be larger to accommodate skating, and large probe pads add parasitics to the measurement. Another drawback to W is that its contact resistance increases with use. With either BeCu or W, the contact pressure has a significant effect on the probe's durability, especially W because of its stiffness. The springy compliance of individual ground and signal probe tips becomes a real advantage when probing nonplanar wafers and packages.

The probes are gold-plated to reduce conductor loss. Nickel under-plating keeps the gold from rubbing off, yielding longer probe life. When biasing the DUT, the probe should be able to withstand high dc current.

3.3   Equivalent Circuit

The role of the coplanar probe is to transition from a coaxial cable to a CPW probe pad. The probe's coaxial connector is hooked to the RF test system while the probe tips touch probe pads on the wafer surface. Tapering the probe's body is the smoothest way to transition the electromagnetic fields from coaxial to CPW mode. The ideal transition has a constant 50-Ω impedance throughout. A practical approach is to design the probe for low insertion loss and return loss, at the same time minimizing crosstalk to the die and unwanted modes. Moding limits the maximum frequency of the probe.

Figure 3.4 shows the equivalent circuit of a signal-ground (S-G) coplanar probe as it makes contact to probe pads on the wafer. Capacitance C_p arises between the signal and ground probe tips, with fringing capacitance C_{PS} and C_{PG} from their outer edges. The ground probe tip and the pad it contacts have nearly the same potential, so the capacitance to ground C_{PG} should be small. C_{PS} originates at the signal probe edge and, without any other ground nearby, should be larger than C_{PG}. The exact values of the capacitances will depend on the substrate, its thickness, and its dielectric properties. Because it has one less ground probe, the ground inductance of a S-G coplanar probe is about twice that of a ground-signal-ground (G-S-G) probe.

The potential of the ground probe G can be different from the potential of the ground on back of the wafer due to via-hole inductance. Potential differences in the grounds bring about a common-mode current flowing

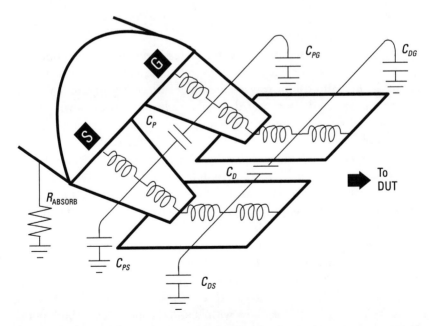

Figure 3.4 Equivalent circuit of an unbalanced coplanar probe contacting probe pads on the wafer.

through the ground probe, impacting C_P and C_D (see Section 8.7.1 in Chapter 8). To hinder common-mode current, an RF absorber, R_{ABSORB}, should be applied to the bottom of the probe. The absorber dampens traveling waves produced by a common-mode current within the probe. These traveling waves move back through the probe toward the VNA.

3.4 Characterizing a Coplanar Probe

As it travels through the probe, the RF signal can degrade for three reasons: reflections, crosstalk, and conductor loss [6]. The most common of these is conductor loss, followed by crosstalk. Crosstalk commonly occurs when two signal probes share a third ground probe in between (S-G-S). Referred to as common-mode inductance, the inductance of the center probe couples energy from one signal probe to the other. Common-mode inductance can be measured by shorting all three probes together and looking at the transmission between signal pins (accounting for the shorting bar's inductance). In practice, inductance in the die's ground plane is often a greater source of crosstalk than common-mode inductance within the probes. When the die's ground path has a large amount of inductance, determining the probe's crosstalk becomes a challenge.

Coupling from the probes to the wafer is different from common-mode inductance within the probe. Crosstalk can develop from the underside of the probe to conductors on the wafer surface (see Figure 3.5). This can cause other conductors near the probe pad to resonate, showing up in the measured data. Pad-to-pad coupling is not nearly as troublesome as pad-to-probe coupling. The taper of the probe tips can create slot line modes between the ground and signal probes that radiate. The absorber on the probe's underside helps prevent coupling between the wafer and probe. Ensuring that there is no ground metalization or other conductors underneath the probes will improve repeatability, otherwise measurements can vary as the probe steps from die to die. The die should be at least 500 μm apart for resonance-free probing.

There are three ways to measure the RF behavior of a coplanar probe. In one method, view the coplanar probe as a coaxial-to-CPW adapter. Most VNAs have a calibration option called adapter removal, used for measuring noninsertible devices (that is, those with identical connectors on both ports) [7]. An adapter is needed to mate identical ports directly together. In this case, the adapter is one of the two coplanar probes.

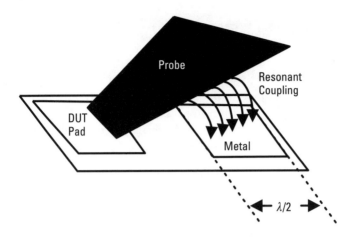

Figure 3.5 Coupling from the probe to adjacent metal on the wafer. Resonance takes place when the adjacent conductor approaches $\lambda/2$ in length.

To illustrate, apply adapter removal to a coplanar probe using a SOLT calibration. Begin by performing the SOL calibration on one probe (see Section 2.6.1 in Chapter 2). The other coplanar probe becomes the "adapter" probe. Next, disconnect the adapter probe and perform the SOL calibration steps at its coaxial cable connector, being sure to enter the VNA calibration coefficients for the coaxial standards. Reconnect the adapter probe and then perform the final thru steps of a SOLT calibration. During the thru measurement, the normal adapter removal procedure calls for swapping the adapter between the two ports. Since coplanar probes are similar (especially new ones), they should not be interchanged. Any unnecessary disconnection and reconnection will contribute calibration error. Once the SOLT calibration is complete, use both coplanar probes to measure the calibration thru standard (being sure to include its calibration offset delay). The test system now considers the "adapter" probe to be the DUT.

A second method of characterizing a coplanar probe uses a SOLR thru calibration rather than SOLT and assumes the ports are symmetrical. Beyond this, the procedure is the same. A third method involves disconnecting both coplanar probes and performing a coaxial two-port calibration, defining the reference plane at the ends of the coaxial cables. Afterwards, reconnect one of the coplanar probes and perform a one-port calibration on the probe. Now reconnect the second probe and measure known elements, such as the calibration standards, to determine the behavior of the second coplanar probe.

3.5   Using Coplanar Probes

3.5.1   Planarization

When a coplanar probe wears out, the tips are often no longer planar (see Figure 3.6). For example, consider the G-S-G probe shown in Figure 2.12 in Chapter 2. Two 100-Ω resistors in parallel produce a 50-Ω load. If only one of the ground probes touches, then the load will appear as 100-Ω. A more subtle point is that the reference plane at the probe tips is no longer in a straight line. Poor reference plane definition will introduce uncertainty in the DUT measurement.

The springy compliance of individual signal and ground probe tips becomes a real advantage when probing nonplanar surfaces on wafers and packages. However, this can be a drawback. Since each probe tip is independently flexible, any reference plane configuration is possible. The best way to check probe tip planarity is with a contact substrate, which is simply a metalized field. When the probes are lowered onto this field, the probe tips leave scratches or "footprints" in the metal (see Figure 3.7). The scratch depth indicates when the probe is tilted or mounted at an angle. Another method for checking probe planarity is by using an ohmmeter. To do this, short the signal and ground probe tips by pressing them to the contact substrate, verifying both probes are touching. If they are not touching, an open

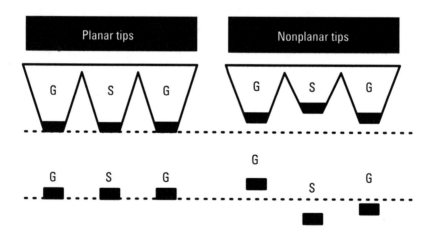

Figure 3.6   Coplanar probe planarization. The dotted line indicates the calibration reference plane. Shown on the right are probes bent after lengthy use. Even if the calibration results look good, a common reference plane has not been established.

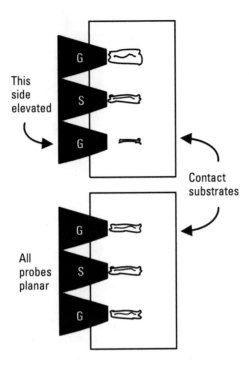

Figure 3.7 Using a contact substrate. In the top figure, light scratches left by the lower probe tip indicate its higher elevation. The figure at the bottom shows even scoring by all probe tips.

will register on the ohmmeter. Of the two, looking at the probe tips through a microscope is preferable to blindly observing its electrical response.

Some probe micropositioners offer a planarity adjustment. If this is not available and it is seen that one side of the probe is higher than the other, then place metal shims under one side of the probe to elevate it.

3.5.2 Alignment

Measuring die on opposite sides of the same wafer requires precise alignment of the wafer and probe. The probe tips should line up to an edge of the die, forming a plane. If the wafer is oriented at an angle to the probes, then moving a distance across the wafer will not land the probe tips on another die's pads.

To align the probes and the die, sweep the wafer chuck from left to right, observing an edge of the die as it scans. If the probe tips are not evenly aligned with the wafer, then the edge will appear to rise and fall. Adjust the theta rotation of the chuck until the die's edge sweeps across the wafer in a straight line. When wafer scanning is not possible, use the outer corners of a die to align. The outer corners of the probes will define the reference plane.

3.5.3 Skating

When a coplanar probe initially touches down on a probe pad, the tip arrives normal to the wafer surface. Yet because the probe's body is at an angle to the wafer, lowering it further causes the tip to glide across the pad. This phenomenon is known as skating. Some skating is always necessary since the signal and ground pad metal may not be deposited with the same thickness on all wafers. In some instances, the surface of a wafer may not be planar, especially when multiple dielectric and metal layers are used in making it. *Chemical-mechanical polishing* (CMP) during IC processing leaves each layer with $+/- 0.5 \mu m$ flatness across the wafer.

When designing the die, alignment marks near the die's pads help to align the coplanar probe for a consistent amount of overtravel. Figure 3.8 shows triangular alignment marks designed for a 1-mil probe skating. The marks ensure the probes stop at an exact spot on the standards that correspond to the calibration coefficients entered into the VNA.

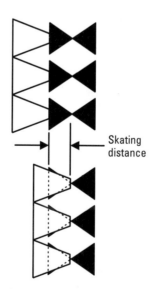

Skating distance

Figure 3.8 Using alignment marks to set skating. Initially, the probes touch down on the outer edge of the triangles (in the top part of the figure). Then adjust the z-axis height until the probe skates across the substrate, halting at the centers of the triangles (in the bottom part of the figure).

Skating has both a mechanical and an electrical impact on the measurement. Mechanically, skating leaves conspicuous marks on the die pad, visible under a microscope. These marks tell when a die has been probed, useful in post-probe inspection. A disadvantage is that the scrubbing action of the skating scrapes some metal off the pads. Over time this accumulates on the probe tips, requiring periodic cleaning. Electrically, pad overtravel resembles an open stub (see Figure 3.9). An open stub attenuates an RF signal at higher frequencies. Different amounts of skating can mistakenly resemble nonlinear behavior in the DUT.

The equivalent circuit of a pad stub is shown in Figure 3.10(a). The interaction of the stub and probe can be quantified by first performing a TRL using an alumina calibration substrate. Once calibrated, measure another thru on the wafer with probe pads. Use offset delay to shift the reference plane to where the probe pad meets the thru. An RF measurement will then quantify the probe pad–to–probe tip electrical model. Finally, fit the measurement to the equivalent circuit shown in Figure 3.10(b).

During calibration, getting the probes to contact the same place on the calibration standards each time is difficult. A common mistake is to readjust the probes when moving from one calibration standard to another. Before calibration has begun, set the micropositioners once and do not change their position during calibration. Moving the probes changes the standards' parasitics, particularly the inductance (see Figure 3.11). With coplanar probes, the amount of inductance depends on the amount the probe overlaps the standard.

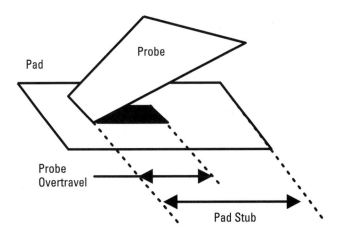

Figure 3.9 Diagram showing probe overtravel and the resulting pad stub length.

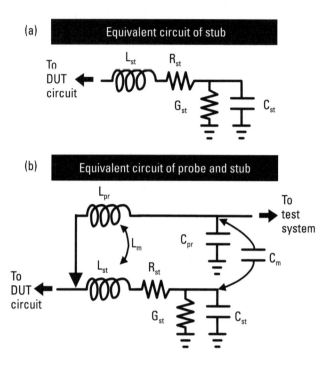

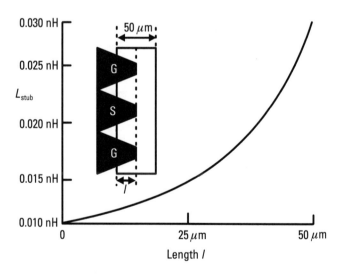

Figure 3.10 (a) Equivalent circuit of stub; and (b) equivalent circuit of probe and stub.

Figure 3.11 Electrical impact of skating on a shorting bar. As the probe skates a length l, it increases the stub's inductance L_{stub}.

In general, too much skating shortens the length of the calibration standards (see Figure 2.23 in Chapter 2). With the thru standard, the reference plane is no longer centered in the middle of the thru. Excess skating makes the thru's delay entered into the VNA too long, causing the reference plane of each port to overlap with the other. To verify the effect of excessive overlap, remeasure the open standard after calibration. The return loss should not be greater than 0 dB. If this is the case, then the load measurement may have had too much skating.

3.5.4 Cleaning

After extended use, probes accumulate debris such as lifting probe pads or metal strings from around the pads. Probes can be cleaned with either compressed air or IPA applied with a swab. The swab should have a sponge tip rather than cotton, which can leave behind threads. Always wipe in a direction away from the probe tip. Be aware that IPA will flow into cavities in the probe. Since the cavities are not directly exposed to air, it can take some time to dry (often 15 minutes or more). As it dries, the RF measurement of an open will change with time.

Another method of cleaning is with an alumina substrate. Lightly bouncing the probe tip onto a smooth, nonmetallized area of the substrate will knock debris off the tip. A gel pack can be used in a similar fashion. A gel pack is a small plastic package with gelatinous, sticky material inside, used to store die or small components. Lowering the probes onto the gel pack will stick the debris to the gel.

A vibrating scrub pad, programmed to scrub a needle probe when a number of bad die are measured in a row, can be found on some probe stations near the wafer chuck. Scrub pads are typically used with needle probes and should be applied judiciously to coplanar probes. The scrubbing action wears out the probes faster.

3.6 Probe Configurations

3.6.1 Balanced

The most popular coplanar probe configuration is the G-S-G probe. Two ground probes shield a single signal probe in between. Its principal advantage is in tightly controlling the fields around the signal probe. Electric fields emanating from the signal (S) probe terminate on the ground (G) probes while the magnetic fields between S and G cancel. For example, consider probing two 100-Ω resistors in parallel that equate to a 50-Ω load (see Figure 3.12). In this case, the mutual inductances M on either side of S are balanced due to equal current division between the grounds.

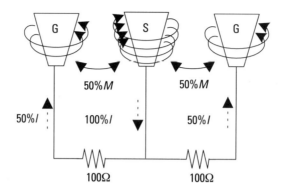

Figure 3.12 Balanced probing a load. The magnetic fields (circular lines with arrows) have equal mutual inductance *M* on either side. The current flow *I* (dotted lines) is evenly split between the grounds.

Now consider the case of an unbalanced load where the ground currents are not evenly split (see Figure 3.13). Such an imbalance might be intended to emulate the current flow in a DUT. While the parallel-equivalent load is still 50Ω, this imbalance introduces small error in the calibration. The probe's parasitics will not equal those in the balanced current case (see Figure 3.12). Since the VNA calibration coefficients are designed to correct for balanced currents, using the load in Figure 3.13 can add error to the calibration. Unbalanced currents give rise to higher-order modes that can cause oscillations in a DUT. To overcome this, the DUT should be redesigned for balanced current loading or the calibration load redesigned to emulate the load presented to the DUT. Differences in the DUT and calibration currents can make a difference when measuring miniscule amounts of capacitance and inductance such as in device modeling.

Compared to G-S-G probes, signal-ground-signal (S-G-S) probes suffer from crosstalk and poor isolation. Inductance in the shared ground probe serves as a coupling path between the two signal probes. To quantify the G probe's inductance, contact the probes to a shorting bar and measure the insertion loss between the two signal probes. In general, S-G-S probes are best used below 10 GHz.

3.6.2 Unbalanced

The principal advantage of an unbalanced probe (G-S or S-G) is in reducing the overall die size. One less pad means a slightly smaller die, thereby increasing the number of die per wafer. The drawback is that one less ground probe

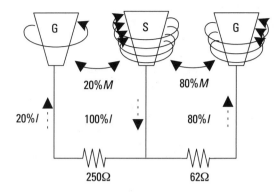

Figure 3.13 Probing an unbalanced load. The magnetic fields (circular lines with arrows) around the right ground probe are stronger than those around the left ground probe, affecting the self and mutual inductances.

means less shielding, resulting in crosstalk problems (see Figure 3.14). A second ground probe on the other side of the signal probe ensures a better-controlled mode of CPW wave propagation along the line as compared to G-S (or slotline) propagation. Even calibrated, the right side of the unbalanced probe shown in the bottom of Figure 3.14 can couple to the substrate, limiting its high-frequency response. Also, resist the temptation to use an unbalanced probe on balanced probe pads. In such a case, the unused ground pad will behave as an RF open stub in the ground path's equivalent circuit.

G-S-G probes can also have an unbalanced design. The distance between probe tips (referred to as the pitch) need not be equal on both sides of the signal probe. RF performance degrades the more unequal the pitch on either side becomes.

3.6.3 Differential

Both balanced and unbalanced probes are considered single-ended, also known as common-mode. In common-mode circuits, the signal voltage is referenced to ground. Conversely, signals on differential lines are referenced to one another. In general, differential circuits exhibit better crosstalk immunity and dynamic range than ground-referenced circuits.

One way to convert a single-ended probe to a differential probe is to install a 180° splitter-combiner balun (see Figure 3.15). The signal will propagate between the pair of lines as opposed to between the signal and ground. In principle, the same calibration techniques apply to differential probes. However, measuring differential S-parameters may require a VNA

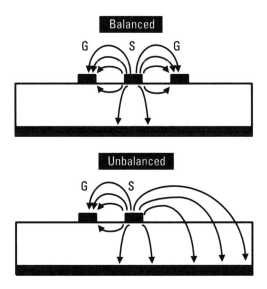

Figure 3.14 Comparing the electric field pattern of a balanced coplanar probe (top figure) to an unbalanced probe (bottom figure). In the unbalanced case, the electric field lines on the right are a source of crosstalk.

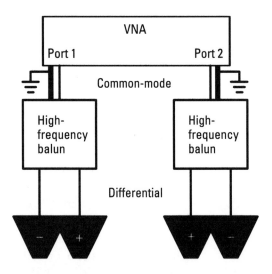

Figure 3.15 Differential probing using a high-frequency balun in-line with a single-ended connection. No ground connection is made to the DUT, rather a virtual ground exists at the center of the differential probe.

with twice as many ports. Performing a calibration requires twice as many standards, and the error model is twice as large [8, 9].

3.6.4 Other Probe Configurations

The dc and RF can be applied to the die using needle probes for the dc and coplanar probes for the RF, often with both probes on the same probe head. Similarly, mixed-signal ICs have both digital and low-frequency RF signals. Instead of S-parameters, digital designers study the pulse propagation delay and the pulse's rise and fall times.

The high-frequency behavior of coplanar probes is far superior to needle probes. Needle probes are commonly used for dc biasing and are reliable for carrying RF signals up to a few megahertz. Because they are long, needle probes have excessive series inductance (~1 nH/mm) and minor fringing capacitance, limiting their high-frequency ability. Needle probes are constructed of a solid conductor with no corresponding ground shielding. High-frequency signals require a ground path as close as possible to the signal path. Simply making a dc ground strap connection to the chuck will not suffice for an RF ground at high frequencies.

3.7 Noncontact Probing

In addition to coplanar probes, there are noncontact methods of probing the die [10, 11]. Noncontact probing is useful for exploring areas of the circuit lacking coplanar probe pads. Mounting a miniature antenna or field probe above the die is one noncontact method. To do this, terminate the unused ports of the DUT and inject a signal into the port of interest. Then use a field probe to detect electromagnetic fields along the exposed DUT circuit, particularly at transmission line discontinuities that exhibit current crowding and standing waves. An electric field probe can be as simple as a coaxial cable whose braided shielding and dielectric is stripped back, revealing the center conductor as the probe [12]. To detect magnetic fields in the microwave range, a quadrupole antenna can be made at the end of an alumina substrate (see Figure 3.16) [13]. Connected to a spectrum analyzer, the antenna measures the amplitude and phase of signals along either a microstrip or a CPW line on the die. The loop's dimensions determine the probe's sensitivity and its center frequency.

For frequencies above 1 GHz, another noncontact probe is the bolometer. Resembling a thin-film resistor, the bolometer is a thin bismuth film deposited between two contacts on a substrate. The contacts connect to an amplifier to boost the received signal. When exposed to microwave radiation

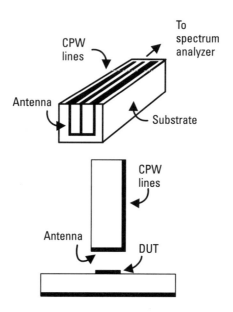

Figure 3.16 Two-loop quadrupole probe. The top figure shows its construction. The bottom shows how to position it over a DUT.

on the surface of a die, the bolometer absorbs energy and heats up. As it heats, its resistance decreases. The voltage across the bolometer is proportional to the power received. Pulsing the RF to the DUT makes it easier to discern a low-level signal on the spectrum analyzer or a microwave transition analyzer [14]. The measured voltage can also be transformed to S-parameters. More complicated noncontact techniques involve probing the DUT surface with electron beams and laser beams.

Noncontact probing works best in a shielded environment free of stray fields. Use a cover plate to both shield the DUT and support the probe a fixed distance above the die. To calibrate a noncontact probe, the probe is placed over a circuit element such as a 50-Ω transmission line, whose voltage is already known.

3.8 Applications

3.8.1 Millimeter-Wave Probing

At millimeter-wave frequencies (greater than 40 GHz), the RF test system connects to a coplanar probe either with a small-diameter coaxial connector or a rectangular waveguide. Since the coplanar probe has a coaxial line running internally through its body (see Figure 3.3), the millimeter-wave coplanar probe can have two transitions, one from rectangular waveguide to coaxial line and another from coaxial line to CPW. In the initial transition, the coaxial center pin can serve as a field probe inside the rectangular waveguide [15]. As the frequency increases above 40 GHz, both coaxial cable and waveguides become lossy. The body of a millimeter-wave coplanar probe tends to be short because lengthy ones have high skin and radiation losses. A solution to high-frequency loss is to perform frequency upconversion by using a mixer mounted on or near the probe [16]. By operating the rest of the test system at a lower frequency, less loss is accrued. Also, mounting an amplifier near the probe-to-DUT interface can help to overcome test system losses.

Here are some miscellaneous points to consider when millimeter-wave probing. When designing the DUT, make sure that transmission line discontinuities (such as bends in the line) are a sufficient distance away from the calibrated reference plane. Transmission line discontinuities can excite evanescent modes. Keeping the discontinuities far enough away from the RF probe makes sure the amplitude of the modes has decayed sufficiently before reaching the calibrated reference plane. Some calibration methods are better than others at dealing with moding [17].

At millimeter-wave frequencies, constructing calibration standards is difficult since the operating wavelength is so small. When a TRL calibration cannot be used, build SOLT standards and then characterize them in a TRL-calibrated system. Use the measurements to find the calibration coefficients of the SOLT standards. Once characterized, the standards can then be put to use as millimeter-wave SOLT standards [18].

3.8.2 Impedance-Matching, Low-Impedance Probes

To impedance-match means to transform the test system's impedance (usually 50Ω) to the DUT's port impedance, minimizing reflections. For instance, measuring highly reflective, low-loss loads is important for *power amplifiers* (PAs). Power amplifiers have output impedances in the $2–10$-Ω range, but are problematic in a 50-Ω test system [19]. A probe not matched to the PA's low output impedance causes reflections and loss at the probe tip-to-PA interface.

The transforming network can be built into a coplanar probe or a circuit mounted nearby. Low-impedance probes also improve dc biasing because a lower probe impedance leads to a smaller voltage drop across the probe.

Often an impedance-matching probe is not available, only a standard 50-Ω probe. To calibrate a 50-Ω probe for a low-impedance DUT, use calibration standards that are the conjugate match of the DUT's impedance. For instance, consider a 50-Ω probe used to measure a 10-j0.5-Ω PA impedance. For minimum reflection, the SOL standards should ideally exhibit a parasitic inductance of j0.5Ω. The load standard should have the same resistance as that of the DUT port, in this case, 10Ω. Upon calibration, the probe will then impedance-transform the measurements to the test system's 50-Ω impedance. In reality, building standards to match the DUT's port impedance is a time-consuming approach.

3.9   Coplanar Probe Calibration Standards

This section details how to characterize calibration standards made for coplanar probing. It builds on the calibration concepts introduced in Chapter 2 and specifically applies them to CPW standards.

Figure 3.17 shows three CPW calibration standards, a thru, an offset short, and an offset open [20]. The physical reference plane is offset from the electrical calibration plane by a length l. The parasitics introduced by the offset line are dealt with using the techniques discussed in Section 2.5. Regarding the parasitics of the standards themselves, these can be viewed as extending the electrical reference plane by an amount l_{ext} beyond l [21]. The offset short has inductance due to the large area of shorting metal at its end. Similarly, electric fields fringing from the end of the offset open behave capacitively. First quantify the parasitics of the standards individually; then adjust the electrical reference plane by an amount l_{ext}. For the offset open, this works as

$$l_{ext} = \frac{C_0}{C'} \qquad (3.1)$$

where C' is the capacitance per unit length of the CPW line and C_0 the parasitic fringing capacitance. C_0 accounts for fringing fields off the end of the CPW open. The value of C_0 depends on the gap width and the CPW line width. Similarly, for the offset short

$$l_{ext} = \frac{L_0}{L'} \qquad (3.2)$$

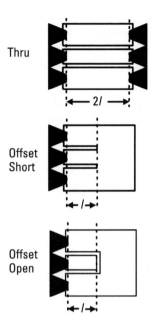

Figure 3.17 Offset CPW calibration standards; *l* is the physical length of the line from the probe tips to the standards.

where L' is due to the inductance of a unit length of the CPW line and L_0 the inductance of the shorting metal at the end of the CPW line. Equation (3.2) can be equally applied to compensate for the pad inductance in a CPW load.

3.9.1 Alumina Calibration Substrate

The most common way to calibrate coplanar probes is with a commercially made calibration substrate. Made of alumina, the substrate holds the standards needed to calibrate using the methods discussed in Section 2.6 (i.e., SOLT, TRL, and LRM).

Employing a calibration substrate can inadvertently highlight the differences in properties between the calibration substrate and the DUT wafer [22]. In general, an alumina substrate exhibits more capacitance than GaAs or silicon wafers because of its dielectric constants. The capacitance of the probe to the alumina substrate will not be the same as the capacitance of the probe to the DUT wafer. Since the calibration and DUT substrates are made of different materials, the pad capacitances of each are different. Calibration coefficients will correct for the pad parasitics as they appear on the calibration substrate, not on the wafer.

A simple way to quantify the differences in parasitics is by making an uncalibrated thru measurement. Using the same probe pad layout on both, measure a thru on the substrate and one on the wafer. The thrus on each should maintain consistent line impedance Z_0 and length. That is, the CPW line on the wafer should be designed with the same Z_0 as on the alumina substrate. Comparing the measurements then gives the difference in pad parasitics.

It is not necessary for all the calibration standards to be on the same substrate. For instance, the thru may be found on the DUT wafer, while the other standards are on alumina. In this way, the thru better resembles the parasitics of the interconnecting lines leading to the DUT. Because of the dielectric constants, the on-wafer thru will have different attenuation and phase characteristics than the alumina standards. This difference can be a source of calibration error.

Recall that when the probe contacts the probe pad, skating leaves an RF stub behind the probe tip (see Figure 3.9). Ideally, the stub on the alumina standard should be the same length as the one left on the DUT's pads. Although the stub lengths may be the same, different dielectric constants contribute different shunt capacitances. To quantify this difference, put the same pad pattern on both the alumina substrate and the wafer. Then measure and compare the shunt capacitances.

3.9.2 On-Wafer Calibration Standards

Rather than on an alumina substrate, the calibration standards can be placed on the same wafer as the DUT. This section discusses the issues that arise when the calibration standards are on-wafer with the DUT.

Calibration methods such as SOLT and LRM base their validity on a precision load standard. The problem with placing the load on-wafer is that IC fabrication processes are not entirely stable. If the doping implant drifts, then the load's impedance will drift along with the DUT's behavior. Under these circumstances, calibrating with an on-wafer load is particularly troublesome. When the IC process drifts, the best on-wafer standards are an open and a thru, such as those used with TRL. Employing TRL on-wafer, the thru compensates for CPW dispersion as well as insertion loss. It can also be used to de-embed the interconnecting lines leading up to the DUT (see Section 6.3 in Chapter 6). Unfortunately, TRL calibration tends to be narrow band. Wider-band calibrations require multiple line lengths that take up valuable space on a wafer. Still, because it normalizes to the characteristics of a line on the wafer, TRL is the preferred on-wafer calibration technique above 5 GHz.

Other IC processes, such as planarization, can also influence the calibration. Planarization regulates the thickness of a dielectric on a wafer, affecting some of the calibration standards. For instance, the via-hole inductance of a short varies by thickness. Compared to the line etch process, the inductance of a via hole has a greater impact on calibration than the metallization etch variance affecting the transmission line's width. The line's width determines its characteristic impedance Z_0. In other words, the planarization process can be more important than the lithographic process that defines the conductor lines.

To quantify comparisons such as these, employ a two-tier calibration (see Section 5.4 in Chapter 5). First perform a coaxial calibration to the ends of the RF coaxial cables connecting to the probes. This de-embeds large variances and irregularities due to the test system. For the second tier, reconnect the probes and calibrate on-wafer using TRL. After calibration, measure the line standard at several die locations across the wafer. The variance in the line's measurement illustrates how the line's properties change across the wafer. When examining planarization, the RF measurements will not vary across a single wafer as much as wafer-to-wafer.

The type of transmission line (such as microstrip or CPW) also plays a part. With microstrip, wafer process variability will impact the line's impedance Z_0 and its propagation constant γ. CPW lines tend to be dispersive, making on-wafer standards placed at the ends of CPW lines also appear dispersive. In CPW, wafer process variability sometimes manifests itself in the form of high-frequency moding.

3.9.3 Open: In the Air or on Open Pads

Suspending coplanar probes in the air is the most common open standard. This sets up a *transverse electromagnetic* (TEM) mode of wave propagation at the probe tips. Simply lifting the coplanar probes into the air suffices as an open. Free space, however, is an uncontrolled environment. A probe can receive radiation and other stray fields much like an antenna (see Section 3.7). Depending on how far the probe is in the air, metal lines on the surface of the wafer can attract stray electric fields from the signal probe tip, increasing loss. To avoid this, coplanar probes should be raised at least 0.010 inch above the wafer surface.

An alternative technique is placing the probe tips on a bare area of the calibration substrate. The substrate's dielectric offers controlled parasitics to de-embed from the RF measurement. Compared to air, repeatability is worse because exact placement on the substrate varies. Landing the probes at a different spot with different dielectric characteristics can produce a negative capacitance measurement. Because the substrate's surface gives rise to quasi-TEM propagation, the substrate open is susceptible to dielectric losses not easily described with the VNA calibration coefficients. After calibrating using an open on the substrate, gain may appear when measuring with the probes up in the air because air loads the probe less than placing it on the wafer.

The open parasitic is modeled as a negative capacitance. This is not due to a "negative" pad inductance, rather it comes from electric fields between the ground and signal probes extending outside the probe's calibrated reference plane (see Figure 2.10 in Chapter 2) [23]. During test, the probes rest on the DUT's probe pads and the capacitance outside the reference plane is not present. Hence, their contribution must be de-embedded from the open standard.

3.9.4   More Design Tips

Here are some more tips on designing better coplanar probe calibration standards. The physical length of the standards should be less than a wavelength long. This keeps undesired modes from propagating, improving calibration accuracy. To ensure the standards launch a CPW mode and not microstrip or parallel plate, choose a thin substrate and do not plate the backside [24].

To permit the wafer chuck to step-and-repeat from standard to standard, evenly space the standards when laying them out. The amount of coupling between the standards depends on how physically close they are. Regardless of the standards' RF behavior, the probes themselves may couple. During calibration, two coplanar probes are typically mounted opposite from one another, contacting separate standards. Fixing the distance between probes ensures the parasitic coupling from probe to probe remains constant [25]. Calibration will remove some, but not all, of such coupling error.

Applied to coplanar probes, the SOLT calibration method has practical advantages. The two grounds on the top surface of a CPW thru are wide enough to accommodate different-size probe pitches. While probes of varying pitch can physically contact the thru and other standards, their calibration coefficients will still depend on that probe's particular pitch. Because the parasitics of the standards correspond to the probe's pitch, dissimilar-pitch probes on opposite ports should be avoided.

During thermal measurements, the load's resistance value changes with temperature, so the calibration substrate is usually held on a detached, thermally controlled area apart from the chuck.

3.9.5 Verification

As discussed in Section 2.7 in Chapter 2, any previously characterized structure can be used to verify a calibration. Rather than using a well-characterized
element, an alternative method employs a resonator designed for low coupling and insertion loss below 50 dB (see Figure 3.18) [26, 27]. With such a
low coupling coefficient, the losses of the test system will not significantly
load the Q of the resonator, making the loaded and unloaded Q's nearly
equal. This decouples the verification element (i.e., the resonator) from the
effects of the test system. This permits the characteristics of the resonator to
be measured with the test system uncalibrated. Use the uncalibrated measurement of the resonator to solve for its propagation constant γ and its effective
relative dielectric constant $\varepsilon_{r,\text{eff}}$. These should be the same γ and $\varepsilon_{r,\text{eff}}$ found
from a transmission line found on the same substrate as the resonator but
measured using a calibrated system.

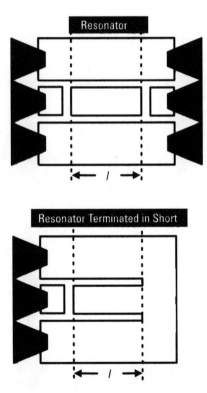

Figure 3.18 Coplanar waveguide resonators serving as verification elements. The resonator length is l.

3.10 Summary

This chapter provides the reader with insight into the design and proper use of RF coplanar probes. Sufficient details are given to develop an understanding of its electrical behavior, how to calibrate and design its calibration standards, and verify the probe's RF performance.

References

[1] Liu, Y., and T. Itoh, "Leakage Phenomena in Multilayered Conductor-Backed Coplanar Waveguides," *IEEE Microwave and Guided Wave Letters*, Vol. 3, No. 11, 1993, pp. 426–427.

[2] Jackson, R., "Mode Conversion Due to Discontinuities in Modified Grounded Coplanar Waveguide," *IEEE Microwave Symposium Digest*, 1988, pp. 203–206.

[3] Leroux, H., et al., "The Implementation of Calibration on Coplanar Waveguide for On-Wafer Measurements at W-Band," *IEEE Microwave Symposium Digest*, 1996, pp. 1395–1398.

[4] Williams, D., et al., "On-Wafer Measurement at Millimeter Wave Frequencies," *IEEE Microwave Symposium Digest*, 1996, pp. 1683–1686.

[5] Carbonero, J., G. Morin, and B. Cabon, "Comparison Between Beryllium-Copper and Tungsten High Frequency Air Coplanar Probes," *IEEE Trans. on Microwave Theory and Techniques*, Vol. 43, No. 12, 1995, pp. 2786–2793.

[6] Carlton, D., K. Gleason, and E. Strid, "Microwave Wafer Probing," *Microwave Journal*, Vol. 28, No. 1, 1985, pp. 121–129.

[7] Hewlett-Packard, "Measuring Noninsertable Devices," *Hewlett-Packard Application Note 8510-13*, 1991.

[8] Bockelman, D., and W. Eisenstadt, "Calibration and Verification of the Pure-Mode Vector Network Analyzer," *IEEE Trans. on Microwave Theory and Techniques*, Vol. 46, No. 7, 1998, pp. 1009–1012.

[9] Bockelman, D., and W. Eisenstadt, "Pure-Mode Network Analyzer for On-Wafer Measurements of Mixed-Mode S-Parameters of Differential Circuits," *IEEE Trans. on Microwave Theory and Techniques*, Vol. 45, No. 7, 1997, pp. 1071–1077.

[10] Lucyszyn, S., et al., "Measurement Techniques for Monolithic Microwave Integrated Circuits," *Electronics & Communications Engineering Journal*, Vol. 27, No. 4, 1994, pp. 69–76.

[11] Dahele, J., and A. Cullen, "Electric Probe Measurements on Microstrip," *IEEE Trans. on Microwave Theory and Techniques*, Vol. 28, No. 7, 1980, pp. 752–755.

[12] Gao, Y., and I. Wolff, "Measurements of Field Distributions and Scattering Parameters in Multiconductor Structures Using an Electric Field Probe," *IEEE Microwave Symposium Digest*, 1997, pp. 1741–1744.

[13] Osofsky, S., and S. Schwarz, "Design and Performance of a Noncontacting Probe for Measurements on High-Frequency Planar Circuits," *IEEE Trans. on Microwave Theory and Techniques*, Vol. 40, No. 8, 1992, pp. 1701–1708.

[14] Wei, C., Y. Tkachenko, and J. Hwang, "Noninvasive Waveform Probing for Nonlinear Network Analysis," *IEEE Microwave Symposium Digest*, 1993, pp. 1347–1350.

[15] Liu, S., and G. Boll, "A New Probe for W-Band On-Wafer Measurements," *IEEE Microwave Symposium Digest*, 1993, pp. 1335–1338.

[16] Yu, R., et al., "Full Two-Port On-Wafer Vector Network Analysis to 120 GHz Using Active Probes," *IEEE Microwave Symposium Digest*, 1993, pp. 1339–1342.

[17] Seguinot, C., et al., "Multimode TRL—A New Concept in Microwave Measurements: Theory and Experimental Verification," *IEEE Trans. on Microwave Theory and Techniques*, Vol. 46, No. 5, 1998, pp. 536–542.

[18] DuFault, M., and A. Sharma, "A Novel Calibration Verification Procedure for Millimeter-Wave Measurements," *IEEE Microwave Symposium Digest*, 1996, pp. 1391–1394.

[19] Basu, S., et al., "Impedance Matching Probes for Wireless Applications," *Cascade Microtech Application Note AR125-0498*, 1998.

[20] Williams, D., and T. Miers, "De-embedding Coplanar Probes with Planar Distributed Standards," *IEEE Trans. on Microwave Theory and Techniques*, Vol. 36, No. 12, 1988, pp. 1876–1880.

[21] Beilenhoff, K., et al., "Open and Short Circuits in Coplanar MMICs," *IEEE Trans. on Microwave Theory and Techniques*, Vol. 41, No. 9, 1993, pp. 1534–1537.

[22] Walker, D., and D. Williams, "Compensation for Geometrical Variations in Coplanar Waveguide Probe-Tip Calibration," *IEEE Microwave and Guided Wave Letters*, Vol. 7, No. 4, 1997, pp. 97–99.

[23] Hewlett-Packard, "Specifying Calibration Standards for the HP 8510 Network Analyzer," *Hewlett-Packard Product Note 8510-5*, 1986.

[24] Gronau, G., and A. Felder, "Coplanar-Waveguide Test Fixture for Characterization of High-Speed Digital Circuits up to 40 Gbs," *Electronics Letters*, Vol. 29, No. 22, 1993, pp. 1939–1941.

[25] Pla, J., W. Struble, and F. Colomb, "On-Wafer Calibration Techniques for Measurement of Microwave Circuits and Devices on Thin Substrates," *IEEE Microwave Symposium Digest*, 1995, pp. 1045–1048.

[26] Woodin, C., and M. Goff, "Verification of MMIC On-Wafer Microstrip TRL Calibration," *IEEE Microwave Symposium Digest*, 1990, p. 1029.

[27] Hettak, K., A. Sheta, and S. Toutain, "A Class of Novel Uniplanar Series Resonators and Their Implementation in Original Applications," *IEEE Trans. on Microwave Theory and Techniques*, Vol. 46, No. 9, 1998, pp. 1270–1276.

4

High-Volume Probing

In years past, wafer testing at the IC foundry was limited to dc needle-probing a few die on each wafer. Simple dc or functional (good-bad) tests were performed using low-speed parametric testers. More robust testing was performed when the die were packaged in their final form. However, weeks or even months could elapse between wafer fabrication and the final packaged test. Any problems with RF performance at the final test often meant scrapping all the wafers in the production line. Throwing away bad die at the final test is costly, since packaging and assembly added value to the components.

The issues surrounding high-volume RF testing are distinctly different from RF probing a single die on the test bench. To illustrate the differences, it is best to describe the environments in which the two operate. Testing in the lab is concentrated in the early stages of an IC design cycle. A test wafer contains a large number of different IC designs. During the wafer test, the chuck is manually moved from design to design. While each design's RF performance is studied in detail, the probes can remain at the same die site for days at a time. On the other hand, high-volume manufacturing concentrates on testing each die as fast as possible. This increases the number of die flowing through a single test system, increasing its utilization and thereby reducing the final cost of a component.

This chapter introduces the reader to high-volume RF probing, focusing on the membrane probe. The membrane probe is the high-volume equivalent of the coplanar probe, serving as the interface from the high-speed RF test system to the die. Since each membrane probe is designed to fit a die's unique pad layout, its RF behavior is related to the die's design.

4.1 High-Volume Test

Using *automated test equipment* (ATE) in high-volume manufacturing shortens the test time and minimizes the number of tests. These two factors determine the test system's throughput, or number of components tested per unit time. High throughput is important since it lowers the final product's cost. Often found in wafer foundries, ATE systems are rack-and-stack stations, made with custom-designed hardware and software. Modern ones are fully integrated, turnkey systems. In general, ATE systems are judged by the following factors [1]:

- Number and quality of the RF ports;
- Number of RF sources;
- RF power supplied to the DUT;
- Types of measurements;
- Flexibility of software;
- Flexibility of DUT test fixture.

An ATE system can be broken into three functional blocks: the RF test equipment, the test head, and the probe card (see Figure 4.1) [2]. Each block can be individually calibrated to ensure system integrity. Typically, RF cables connect the RF test equipment to the test head (sometimes called the *docking head*). The test head connects to the RF probe card either directly or via a cabling assembly. This chapter will first address the RF probe card.

4.2 RF Probe Card

While conceptually a single unit, the RF probe card is physically composed of two pieces, an outer load board and an inner probe board. The load board is usually a PCB, routing connections from the test head to the probe board. The probe board is replaceable, so that worn ones are easily swapped out without replacing the load board.

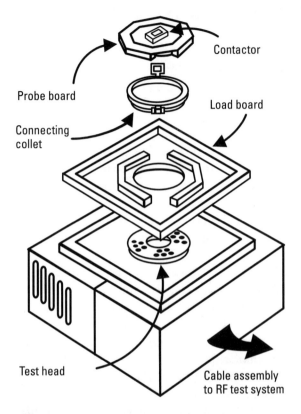

Figure 4.1 A high-volume test assembly. The contactor, probe board, its connecting collet, and load board comprise the RF probe card.

4.2.1 Load Board

Physically, the load board is either a PCB or a loose assembly of cables that connects RF and dc to the probe board. Some load boards use blind-mate connectors, permitting off-the-shelf swapping to suit the DUT style. A guide-and-locking mechanism ensures proper alignment. Rather than a PCB, a cabling assembly offers the flexibility to quickly reconfigure the test system for other style DUTs. Because the cabling can be connected in a number of ways, the disadvantage of using a cabling assembly is that it requires trained personnel to connect it.

An advantage of the load board is that it can contain supporting components needed to customize the test setup to the DUT. It can also hold custom circuitry such as bias tees, relays and switches, terminations, filters, or amplifiers.

4.2.2   Probe Board

Connected to the load board, the probe board is a custom-designed PCB that routes signals to the contactor. A key element of the probe board is the contact area or contactor, since it physically and electrically contacts the DUT (see Section 4.2.3). How well the probe board is designed impacts the test system's RF performance, the test yield, and the cost of the test. Good probe boards furnish a high pin count and are capable of 100,000 contact cycles or more. Careful thought should be given to the probe board's design.

Despite their importance, the probe board and accompanying contactor are often the last parts of the product design cycle to be considered. Usually the DUT's layout is fixed by the time the probe board is given attention. Yet the probe board's behavior (its parasitics, isolation, and line impedance) can limit measurement of the product's RF performance. A product whose performance cannot be reliably measured on a high-volume test system has little commercial value. All the effort that goes into designing an RF product can be lost if a poorly designed probe board is used.

Probe boards often use needle probes for dc and coplanar probes for RF. As mentioned in Section 3.6.4, needle probes tend to exhibit high inductance in both the signal and ground paths. With needle probes, RF bypass capacitors cannot be placed near the DUT. The density of needles around the DUT becomes limited by the size of the needle's points. On the other hand, a high density of probes can be achieved with a membrane probe board. Membrane probes incorporate microstrip lines of nearly any impedance and enclose the ground plane to lower its inductance. This enables the membrane probe to present both a low impedance and a high pad density to the DUT.

4.2.3   Contactor

Attached to the probe board is a contactor containing pads or pins to make physical contact to the DUT's leads (see Figure 4.2). A contactor's key characteristics are repeatability of the contacts, the number of contacts before replacement is needed, and how well it accommodates a range of die or packages.

For leaded packages, the contactor pads are PCB traces, while *leadless chip carrier* (LCC) components use compliant pins. When a pin is bent or the package is not planar, good contact will not be made. Compliancy for each individual pin rather than a row of pins is the best way to ensure contact. The style of contactor pin can be chosen from a number of commercially available spring-loaded contacts, sockets, and elastomers. Pins are often

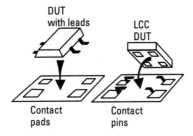

Figure 4.2 The contactor's design should fit the style of the DUT's package leads.

made of BeCu, a flexible metal. With compliant contacts, the amount of travel of the pins within the contactor should be limited. This way, the component will not require precision alignment to gain a repeatable RF measurement.

Some DUT package styles can present challenges. When the DUT is not oriented properly in the fixture, a form-fitted, compliant socket helps keep the package aligned (see Figure 4.3). In addition, the contactor should be replaced as it wears out. Applying even pressure across the surface of the DUT becomes difficult with an ever-increasing number of pads. If the DUT's pads have tight spacing, then it is important to keep the contactor clean to prevent misalignment. Handling becomes tedious as the pad pitch and pad size decreases.

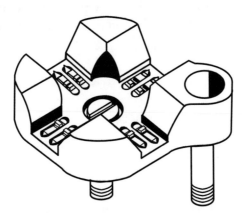

Figure 4.3 A compliant contactor socket for mounting onto a probe board. Each spring-loaded contact in the contactor is individually replaceable. When the contactor body wears out, unscrew and replace.

If the DUT's signal and ground pads are coplanar to one another, then using coplanar probes is an option. However, it is a delicate procedure. Coplanar probing tends to lift one corner of the package as the probe tends to apply more pressure to one side or the other. Unlike a wafer, packaged components are not evenly spaced for assembly-line testing. In practice, loose components are transported from station to station in a tray and are randomly oriented. This makes automated step-and-repeat measurement impossible.

4.3   Membrane Probe

The membrane probe is a new style of RF probe, offering advantages over coplanar waveguide probes along with a few drawbacks (see Table 4.1). This section discusses the construction, design, and proper use of membrane probes.

4.3.1   Construction

The membrane probe is constructed of layers of thin-film polymide pulled over a hard plastic plunger (see Figure 4.4). The flexible polymide is the membrane, conforming the bumps to the DUT surface and enabling one million or more contact cycles. Behind it is the plunger, providing the force for the nickel bumps to contact the probe pads. The bump pattern corresponds to the DUT pad layout. The bumps can contact a DUT with pads as

Table 4.1
Comparing the CPW Probe to the Membrane Probe

Property	CPW	Membrane
Impedance range	5–50Ω	1–100Ω
Number of contacts	2–12	6–1,000
Frequency	dc–110 GHz	dc–20 GHz
Insertion loss at 20 GHz	0.2 dB	2–3 dB
Contact lifetime	500,000	1,000,000
Temperature limit	200°C	125°C
Contact compliance	25 μm	10 μm

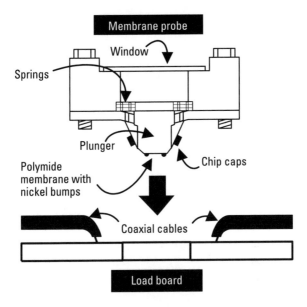

Figure 4.4 Cross-section of a membrane probe about to be affixed to a load board.

small as 50 μm by 50 μm. For probing aluminum pads, the bumps have a cantilever action to scrub through aluminum oxide.

In between the membrane's polymide layers are metallized, conductive planes. Typical membranes have two or three metal layers, one for RF, one for digital, and one for the ground plane. The plane closest to the die is fashioned into RF microstrip lines of characteristic impedance Z_0. These microstrip lines connect to the nonoxidizing, nickel-alloy contact bumps. For high-volume testing, the membrane can be made large enough to contact several die at once. In this way, several DUTs can be measured in one step. Incorporating a wafer edge sensor can also be beneficial. The edge sensor is a simple contact switch, normally open, that detects the edge of a wafer.

4.4 Designing Membrane Probes

As mentioned in Section 4.3, the membrane probe is designed to a specific die layout. Because it is not reconfigurable, it must be changed with each new style of DUT. This section gives tips on designing membrane probes. The ports of the membrane probe can be RF, digital, or dc. Techniques to embed capacitors and inductors in the membrane are also covered.

4.4.1 Digital and RF Signals

This section discusses techniques to improve a membrane probe's digital and RF performance. Some applications call for embedding lumped elements in the membrane. In digital circuits, ripples or ringing on a voltage pulse can lead to bit error. Adding capacitors to the membrane and using low-inductance lines helps prevent this. The closer the capacitors are to the DUT, the more they dampen ringing. Spiral inductors and thin-film or discrete capacitors can be embedded in the thin-film polymide.

For RF testing, care should be taken in routing the RF transmission lines through the membrane. A good transmission line layout has little cross-talk and minimal line loss. High-aspect-ratio lines (line width/polymide layer thickness) exhibit tight field coupling, which yields good isolation [3]. Depositing a ground mesh on one side of a polymide layer and a wide RF line on the other side creates a high aspect ratio. Even so, at high frequencies it becomes difficult to RF-isolate one transmission line from another within the membrane.

RF probing sometimes calls for adding impedance-matching circuits at the test ports [4]. To add inductance to a test port, simply lengthen the RF line within the membrane. Duplicating wire bond lengths found in a package is another reason to add inductance. Shortening or lengthening lines to create RF delay is another. For mixed-signal applications, separating the digital and RF ground planes isolates one from the other. To terminate the DUT's ports, embed a 50-Ω load. Amplifiers can also be incorporated in the membrane to overcome probe board losses [5].

4.4.2 Grounding

Easily overlooked in the design process, the ground plane can have a significant impact on the membrane's RF performance. The ground plane (either meshed or solid) is deposited on one side of a polymide layer. Using a mesh allows the polymide layers on either side of the ground plane to adhere through the holes in the mesh. With respect to the RF transmission line's impedance, holes in the mesh mean less ground can couple to the RF line. A meshed ground plane reduces the capacitance of a microstrip line, while a solid ground plane makes the line more inductive. This feature can be used to tune an RF line for a specific performance. For a given characteristic impedance Z_0, a line with a meshed ground plane is wider than one with a solid ground plane. Wider lines can carry more current, important in high-power applications.

For power and noise testing, grounding techniques can alleviate radiation problems. RF emissions from the top of the die can couple to lines within the membrane, degrading its isolation. A grounded border in the membrane outlining the die helps shield the die from radiation. RF radiation coming from the surface of the DUT or from the membrane surface can also couple to adjacent die, leading to DUT oscillations. To prevent this, adjacent die should be grounded during testing. When grounding multiple points, it is important to ensure that ground loops are not formed within the membrane.

4.5 Using Membrane Probes

A distinct advantage of membrane probes is that ground and signal probe pads no longer have to be coplanar. Instead, the ground and signal pads can be anywhere on the die, offering the utmost in layout flexibility and size compactness (see Figure 4.5). Ground and signal lines no longer need to be brought out to the edge of the die as with coplanar probing, saving die area. Smaller die leads to more die per wafer, increasing wafer yield. However, without a good understanding of electromagnetics, probe pad flexibility can be a pitfall. When the ground and signal are no longer along the same plane, they will not launch a coplanar wave. Using commercially available software to simulate the electromagnetic field behavior of a noncoplanar launch helps to understand its RF behavior.

Planarity affects the lifetime of the membrane probe. If the membrane is not planar with the wafer, then one edge of the bump will wear out faster than the other. Membrane probes touch down normal to the surface of the wafer so that no skating occurs. Without skating, the membrane bumps will contact the same spot on the die's pad each time, resulting in better repeatability. The bumps do not scrub the pads, so die metal should not collect as it does with coplanar probes. No skating also means large probe pads are no longer necessary, supporting higher pad density. To validate a research and development (R&D) membrane design, membranes can even be mounted to the end of a coplanar probe [6].

The following are some ways to improve one's membrane probing technique. Avoid membrane probing lead-solder bumps like those found on solderable package pads. Oxide buildup on lead-solder pads can accumulate on the membrane's RF probe bumps, increasing its contact resistance. Place alignment marks on both the membrane and wafer to make it easier to align the membrane probe to the wafer. Rather than using arrows, as seen in

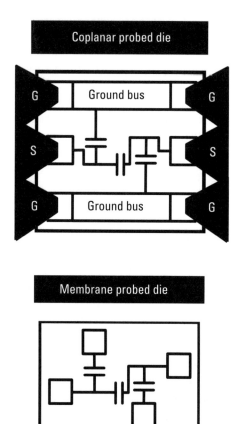

Figure 4.5 Comparing the same filter die laid out for coplanar probing (top figure) and membrane probing (bottom figure). Using membrane probes results in a smaller die size.

Figure 3.8 in Chapter 3, align shapes on the membrane to shapes on the DUT. For example, use parallel bars on both the DUT and membrane. When the two sets of bars align with one another, no light is visible between them. Another alignment technique is to put a border on the membrane that outlines the die's location.

Some rules apply to both coplanar and membrane probes. For example, contacting the pads while the bias and RF are applied (called hot probing) should be avoided, as it can deliver a voltage spike to the DUT and reduce the life of a probe. Use a current-limiting resistor in the bias line to prevent this.

4.6 Calibration

Calibration is one of the most difficult aspects of membrane probes. Because the die's probe pads are unique to the design (see bottom of Figure 4.5), an off-the-shelf calibration substrate is not available. The calibration substrate must match the location of the membrane's bumps. The calibration coefficients entered into the VNA compound the problem. Recall that the VNA coefficients correct for inaccuracies in the calibration standards. Generally, the magnitude of the inaccuracies depends on how far apart the ground and signal pads are. Because an infinite number of pad locations is possible, a generic set of VNA calibration coefficients cannot be used. Instead, the calibration coefficients are determined uniquely to the probe pad layout.

Developing calibration standards and the accompanying coefficients is not a trivial task. The quality of the standards often determines the accuracy of the measurement. Even when the standards and their coefficients are known, calibration may still not be easy. It may be necessary to perform a coaxial calibration and then use time-domain techniques to shift the electrical reference plane to the membrane bumps.

In lieu of a rigorous vector calibration, a scalar calibration may suffice. After a coaxial calibration, try measuring a thru of known insertion loss. Subtracting the thru's insertion loss from measurement gives the insertion loss in the membrane. This moves the reference plane in magnitude but not in phase.

4.7 Summary

This chapter introduces the reader to RF probing for high-volume manufacturing and discusses a distinct set of RF test issues to consider. By gaining an early insight into the testing challenges, a designer can make a product that is better designed for high-volume RF testing, thereby enabling its quick entry into the marketplace.

References

[1]   Gahagan, D., "RF (Gigahertz) ATE Production Testing On Wafer: Options and Tradeoffs," *Cascade Microtech Application Note*, 1999, pp. 388–395.

[2]   Agilent Technologies, *Agilent 84000 RFIC Series Test Systems Product Overview*, 2000, pp. 1–16.

[3]   Cascade Microtech Newsletter, "Testing an 8-Bit, 2 Gigasample Per Second Analog to Digital Converter," *Microprobe Update*, Vol. 28, 1996, p. 3.

[4]   Basu, S., P. Nussbaumer, and E. Strid, "A Membrane Probe for Testing High Power Amplifiers at mm-Wave Frequencies," *Cascade Microtech Application Note 13*, 1998.

[5]   Leung, J., et al., "Active Substrate Membrane Probe Card," *IEEE Electron Devices Meeting Digest*, 1995, pp. 709–712.

[6]   Basu, S., and R. Gleason, "A Membrane Quadrant Probe for R&D Applications," *IEEE Microwave Symposium Digest*, 1997, pp. 1671–1673.

5

Test Fixtures

When the RF test system and the DUT both have coaxial connectors, hooking them together to make a measurement is easy. However, as RFICs have shrunk in size, traditional coaxial connectors are no longer useful. With the multitude of RF component sizes and shapes found on the market, no single connector type fits them all. A test fixture is needed to act as an electrical and mechanical interface from the RF test system to the DUT.

This chapter describes how to design RF test fixtures to measure RF/microwave die and their packages. It covers the fundamental construction of test fixtures, their RF transitions, and how to calibrate them.

5.1   The Basic Test Fixture

An RF test fixture has three basic components, the RF launcher (usually a coaxial connector), RF interconnecting lines leading to the DUT (usually microstrip), and a block housing for the test fixture's body. Figure 5.1 shows a test fixture designed for evaluating leaded packages. Also known as a split block fixture, it consists of separate end sections (A-C and C-A) and a midsection (C-C). Each end section holds an RF launcher and an interconnecting RF line. The interconnecting lines can take any number of forms (i.e., microstrip, stripline, and CPW). The width of the midsection is sized to fit the DUT.

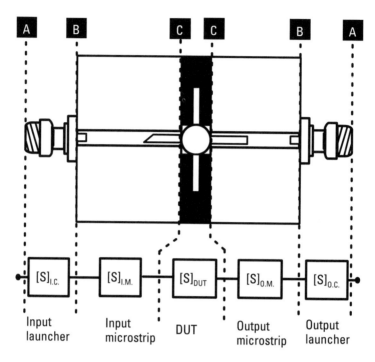

Figure 5.1 A simple test fixture. The letters A, B, and C designate the physical reference planes. S-parameters [S] approximate the RF behavior of each section.

Each section of the test fixture can be individually characterized by its S-parameters and then cascaded for the complete in-fixture response. To do this, convert the [S] matrices to ABCD matrices and multiply them together in a chain fashion [1]. Be sure to consider discontinuities at the interfaces, since these generate reflections. For instance, a coaxial center pin making pressure contact to the microstrip line at B is prone to wear and bending. After extended use, it degrades, causing larger reflections from B. Also, any gap between the coaxial launcher and the fixture block at B results in a discontinuity.

Mechanically, the most useful test fixtures are designed with a modular approach. The fixture block holds the RF launchers and a replaceable midsection while the midsection holds the DUT. The launchers and the block are reusable, while the midsection is customized to fit the DUT. The launchers have alignment pins or bolts that run the length of the block's body. Either spring-loaded end sections or tightening the bolt secures the three sections together.

To calibrate the fixture to the C reference planes, the midsection should be replaced with calibration standards, preferably using SOLT below 5 GHz and TRL above. The parasitics in SOLT standards are difficult to quantify above 5 GHz. The benefit of TRL is that it relies on the characteristic impedance of the line to overcome reflections at the transitions. Because TRL uses different length lines, the width of the midsection (C-C) must be adjustable or the midsections replaceable. Over time, repeated use eventually wears out the calibration standards and launchers. RF repeatability of the launchers will affect the quality of the measurement, particularly at higher frequencies [2].

5.1.1 Qualities of a Good Test Fixture

The qualities of a good test fixture include mechanical durability, good RF performance, low cost, and ease of use. Calibration should be fast and accurate. An ideal fixture would have no loss, a flat frequency response, a linear phase response, no impedance mismatches, a well-known electrical length, total isolation between ports, and be free of external parasitics [3]. Under such ideal conditions, calibration would not be necessary. Most test fixtures offer a tradeoff of these features.

In reality, mounting the DUT into a test fixture clouds the DUT's response, requiring de-embedding of the fixture's effects. Users often compensate for the test fixture through simple means such as calibrating to the ends of the coaxial test cables, subtracting a few tenths of a decibel from the measurement, or altogether ignoring the fixture's effects.

Adhering to a few guiding principles can better improve the result. In general, the RF performance of the test fixture should surpass that of the DUT. For instance, the test fixture's insertion loss should be less than the uncertainty found in the DUT's insertion loss (or gain). All along the RF path, impedance discontinuities at fixture transitions should be minimized. The bandwidth of the fixture should be wider than that of the DUT. Wideband measurements are useful for studying harmonics, for instance. In order to test sensitive devices such as filters and low-noise amplifiers, the fixture's port-to-port isolation should excel.

5.1.2 Characterizing the Fixture's Parasitic Effects

To characterize the test fixture, three approaches are available: create a circuit model of the combined fixture and DUT; create a circuit model of the fixture alone; or de-embed the fixture from the measurement to yield the DUT. The first method requires the DUT to be inserted during characterization.

The measurements are used to create RF models of the fixture and the DUT, generating both models at the same time [4]. Since the reference plane between the fixture and DUT is undefined, this approach can give misleading results. Small parasitics near the DUT-to-fixture interface can often be attributed to either the DUT or the fixture side of the reference plane, a source of uncertainty. The solution is to develop an electrical model of the fixture only, using it to de-embed the fixture from the DUT measurement. Accuracy of the final DUT data will depend on how well the model actually matches the fixture. The third method is to calibrate out the fixture. Mounting calibration standards in place of the DUT shifts the electrical reference plane to the DUT contacts, de-embedding the fixture's parasitic effects. The remaining sections of this chapter detail the RF issues involved in carrying out these approaches.

5.1.3   Types of Fixtures: R&D or Manufacturing

The manufacturing and R&D environments call for different types of test fixtures. Manufacturing demands high throughput, quick insertion, alignment, and securing test fixtures. Rugged, compliant contacts are made to withstand thousands of insertions and removals. The contact area (or contactor) must be kept clean for repeated insertions and extractions. For automated loading and unloading, the manufacturing test fixture tends to be mechanically complicated, providing fast, repeatable connections with acceptable RF performance and long fixture lifetimes.

Conversely, R&D fixtures are unsophisticated and somewhat fragile since only a handful of devices are tested. As shown in Figure 5.1, they can be as simple as a PCB with soldered RF connectors to either end. In R&D testing, the quality of the measurement and how the reference plane is defined are of utmost importance. Correlation and repeatability between manufacturing and R&D test fixtures are often problems.

5.1.4   Fixturing for Passive Components

Passive devices tend to exhibit some loss, so the test fixture's loss must not degrade the sensitivity of the measurement. Unlike active components, passive components have no gain to overcome fixture losses. A good example of the fixture's impact on a passive component measurement is a filter. The test fixture's parasitics can affect the filter's poles and zeroes, shifting its frequency band response. Excessive capacitance or inductance in the fixture easily degrades the filter's band edges, or skirts. For high-reject filters, where only a small amount of signal passes through, port-to-port measurements can be corrupted by fixture leakage.

To reject out-of-band frequencies, the filter presents a highly reflective load to the test fixture port. Rather than RF tuning the fixture's port to match the filter's impedance to 50Ω, instead try placing attenuators (or pads) in the RF path. These should be mounted close to the DUT so that reflected signals are attenuated as soon as they reflect off the filter. Keep in mind that placing attenuators in-line will decrease the dynamic range of the measurement.

5.1.5 Fixturing for Active Components

Designing the best fixture to test an active component depends on the RF application. For instance, an amplifier can easily overcome a test fixture's loss by adding a few decibels of gain to its RF output. However, depending on the number of watts, high-power components can generate a lot of heat. Heat can be dissipated either by providing cooling to the fixture or by pulsing the bias so that only a little heat is generated. Pulsing at a low-duty cycle means the device is biased only for a short period of time, allowing it to cool. When measuring for noise figure or small capacitances (on the order of femptofarads), the DUT should be shielded from any stray fields that can impinge on it. Such sensitive DUTs are *low-noise amplifiers* (LNAs), Schottky diode capacitance, and a transistor's junction capacitances.

Fixture mismatches to the DUT continue to exist even after calibration. The mismatches are mathematically de-embedded from the measurement, not physically removed. After calibration, they will continue to interact with the DUT, especially with reflective DUTs such as filters, LNAs, and PAs. Embedding impedance-matching networks in the fixture can lessen reflections from highly mismatched loads. Active devices such as PAs have a low output port impedance. Similarly, some diodes exhibit a negative resistance. To ensure stability, the loss of the fixture itself should be small but slightly higher than the DUT.

5.2 RF Transitions

RF transitions are key parts of the test fixture, converting the RF signal from one medium (i.e., coaxial, microstrip, and coplanar waveguide) to another. RF transitions perform two tasks. First, an impedance transformation takes place from one characteristic impedance Z_0 to another. Second, it transforms the electromagnetic fields from one transmission mode to another. Carefully designing to these two criteria assures a successful transition.

When a transition is not smooth, reflections and loss appear, caused by sudden changes either in impedance or in the line's geometry. Either disturbs the electromagnetic field pattern. When designed carefully, a transition will not induce reflections. For example, transformers are a simple way to adapt from one resistance to another. Using a 3:2 transformer converts a resistance from 75Ω to 50Ω. Unfortunately, impedance transformations involve both resistance and reactance, not a simple task due to parasitic inductances and capacitances at higher frequencies.

This section details the kind of RF transitions found in test fixtures. When conceptually designing transitions, keep in mind the changing nature of the fields as they transform from one medium to another. The input and output impedances define the end goals.

5.2.1 Coaxial to Microstrip

The most common transition is coaxial to microstrip. Many test systems use coaxial cables, while many test fixtures use microstrip line leading to the DUT. To illustrate the transition, Figure 5.2 shows a coaxial pin protruding through a fixture wall, coming to rest on top of a microstrip line. Ideally, the

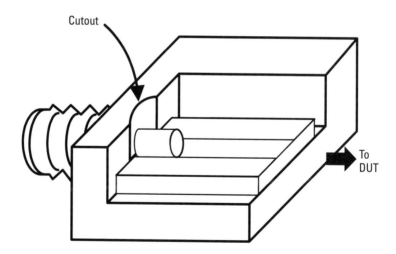

Figure 5.2 Coaxial-to-microstrip transition through the wall of a test fixture. To impedance match the transition, a semicircular cutout in the inner wall can be employed.

microstrip line width corresponds to the diameter of the coaxial center conductor. An oversized-diameter pin will exhibit excess fringing capacitance, limiting its high-frequency operation. A taper in the microstrip line can adequately transition between the two widths. An abrupt change in width from coaxial pin to microstrip line results in a capacitive discontinuity. Using a short-length coaxial pin minimizes the length of the transition. Excess inductance, such as a bond wire or a long length of microstrip line, is a way to offset excess capacitance.

Figure 5.3 offers a detailed look at a the fixture transition [5]. The coaxial connector launches a coaxial TEM mode with the teflon dielectric and center pin's diameter producing a 50-Ω impedance. Conversely, microstrip propagates a quasi-TEM mode. To transition between the two, first ensure the dielectric constants of the two mediums are close. At the fixture wall, the dielectric near the pin suddenly converts from teflon to air, at the same time reducing in diameter. This nearly doubles its characteristic impedance Z_0. Carving a cavity in the wall helps to slowly transform the impedance to 50Ω. Clamping the carrier to the fixture secures a ground connection on the bottom. Applying a conductive, flexible material to the clamp makes certain that pressure is spread equally along the carrier, overcoming poor flatness and improving repeatability. The air gap can induce high-frequency resonances. Keeping the fixture gap small also ensures that the location of the

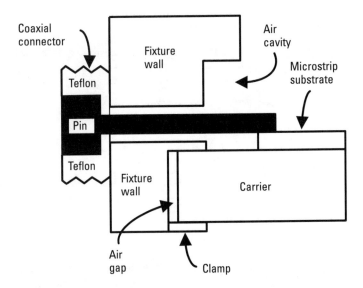

Figure 5.3 Detailed coaxial-to-microstrip transition through the wall of a test fixture.

calibrated reference plane remains the same. To electrically locate the reference plane, short the coaxial pin to the clamp.

Another transition used to gradually offset the fields is the Eisenhart launcher (see Figure 5.4) [6]. In addition to transitioning from coaxial to microstrip, it is equally useful for coaxial-to-CPW transitions when the outer diameter of the coaxial shield corresponds to the spacing of the CPW ground planes.

5.2.2 CPW to Microstrip

This section begins with a poorly designed yet commonly found CPW-to-microstrip transition (see Figure 5.5). Its problem is understood by studying its electromagnetic field patterns. According to electromagnetic theory, CPW launches a vertically polarized (or horizontal) electric field, while microstrip launches a horizontally polarized (or vertical) electric field. At the transition, the ground bond wires form an abrupt "wall" to a CPW traveling wave, undesirably reflecting high-frequency energy [7]. The ground bond wire's inductance creates a potential difference between the microstrip ground and the CPW ground planes. Figure 5.6 shows a better-designed transition. Tapering the grounds avoids an abrupt transition from CPW to microstrip. Various techniques exist for properly designing the taper (i.e., Klopfstein, exponential, and linear).

5.2.3 Rectangular Waveguide to CPW

Generally speaking, rectangular waveguide is best for microwave and millimeter-wave work (greater than 40 GHz). As the frequency increases, it

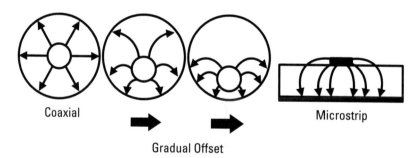

Coaxial   Microstrip

Gradual Offset

Figure 5.4 Transitioning an electric field pattern from coaxial to microstrip using an Eisenhart launcher. The coaxial shield connects to the microstrip's ground plane, while the coaxial center pin rests on the microstrip line.

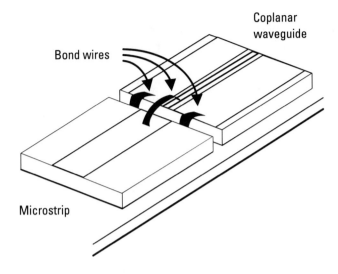

Figure 5.5 A poorly designed CPW-to-microstrip transition, causing reflections at the bond wires.

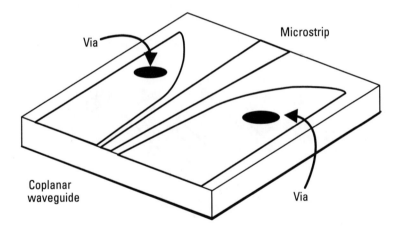

Figure 5.6 Using a tapered ground improves the RF performance of the CPW-to-microstrip transition.

propagates millimeter waves better than coaxial line or CPW. The characteristic impedance Z_0 of rectangular waveguide ranges from 400Ω to 520Ω, while CPW is usually designed for 50Ω [8, 9]. Finline is often used to transition from rectangular waveguide to CPW, reshaping the transverse electric

(TE_{10}) wave to CPW (see Figure 5.7). Normally a wavelength long, the finline's taper splits and rotates the TE_{10} field to conform to CPW. Be aware that finline transitions often have an in-band resonance to avoid. When the DUT requires bias, a better choice is ridge-trough waveguide [10, 11]. It provides a lower cutoff frequency and wider bandwidth than finline due to the capacitance of the ridge.

When designing a waveguide, physically scaling the dimensions to a larger wavelength makes it easier to build. A large-scale version operates at a lower frequency but is easier to debug and test. At the same time, it emulates the field pattern of a higher frequency design.

At higher frequencies (75–110 GHz), radiation becomes the dominant CPW loss mechanism, making calibration problematic. The loss increases with the CPW slot width and signal line width. In such cases, choosing a low-loss dielectric substrate helps.

5.2.4   Rectangular Waveguide to Microstrip

In addition to coplanar waveguide, a finline can transition from rectangular waveguide to microstrip [12]. In Figure 5.7, visualize mounting the microstrip upside down on top of the rectangular waveguide. The finline would

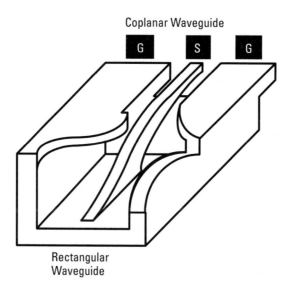

Figure 5.7 A finline transition from rectangular waveguide to CPW. On top, the CPW grounds contact the waveguide walls while the signal line contacts the fin.

contact the microstrip line while the microstrip ground plane forms the top lid, connecting to the outer walls of the rectangular waveguide.

5.2.5 Rectangular Waveguide to Coaxial

Both rectangular waveguide and small-diameter coaxial line can operate in the 40 to 50 GHz range. Transitioning between the two is made possible by inserting a field probe, either electric or magnetic, into the waveguide. When inserted through the waveguide wall, a coaxial cable's center pin can behave as a monopole antenna inside the waveguide. For best results, place the probe at the point of maximum electric field, usually a quarter wavelength from the shorted end of the waveguide. The impedance of the transition is determined by the pin's position and its depth relative to the back wall of the waveguide, often referred to as the back short [13]. Tune the transition by observing the one-port S-parameter, adjusting the probe for minimum return loss. To form a magnetic-field probe, insert a coaxial center pin into the waveguide and loop it to form a round antenna, tuning for the maximum H-field by adjusting the distance of the loop to the back short [14].

5.3 Defining the Reference Planes

A test fixture and DUT can be sectioned into reference planes, as shown by the dotted lines in Figure 5.8. During test, the DUT's leads are bonded to the microstrip between C'-C and C-C'. Because the leads overlay the microstrip,

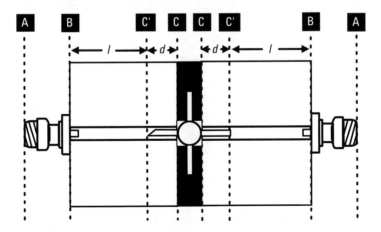

Figure 5.8 Defining the reference planes when measuring a leaded package.

precise definition of the DUT reference plane comes into question. The leads (each having a length d) can be included as part of either the microstrip (B-C and C-B planes) or the DUT (between C'-C'). While the difference may seem subtle, defining the electrical reference planes impacts the DUT's equivalent-circuit model. The leads of the DUT can be partitioned one of three ways.

The first way is to consider the DUT as leadless, including the leads as part of the microstrip lines. The electrical reference plane of the DUT would be at the edge of the microstrip (C'-C'), viewing the DUT as lead-less. The physical and electrical reference planes would lie at the C planes. However, the microstrip lines (C'-C and C-C') are loaded microstrip. Once the DUT is inserted, its leads electrically perturb the d line lengths. This perturbation changes the RF response of the microstrip line lengths directly beneath the leads, invalidating a calibration made at C. Poor reference plane definition is a source of confusion and, ultimately, measurement error.

The test fixture can be calibrated to the ends of the microstrip lines simply by replacing the DUT with calibration standards. Calibration using the TRL method is advantageous since the standards are leadless. One way to deal with the leads is to package SOLT calibration standards (short, open, load, and thru) with leads similar to the DUT. In this way, the standards' leads will affect the microstrip in the same fashion as the DUT's leads. When devising VNA calibration coefficients for the leaded standards, view them as leadless. That is, do not consider the effects of the leads as part of the standards' RF behavior. Once SOLT calibration is complete, the effects of the DUT leads are then considered part of the microstrip between B-C and C-B planes, which is calibrated out.

The second way is to include the leads as part of the DUT but not the microstrip lines to which they are bonded (see Figure 5.9). The microstrip is not part of the DUT but instead is outside the DUT's reference plane. The reference plane outlines the leads of the DUT. Looking at Figure 5.9, it extends vertically through the fixture at C, runs horizontally around the leads on the microstrip surface, and then back vertically across the C plane. The DUT acts as though it were floating in free space, unaffected by the test fixture environment. This would seem the ideal case since the DUT's final electrical model could then be applied to any style of RF board the designer chooses. In reality, attaching the DUT to an RF line will always affect the RF response of the line and the DUT. These electrical effects will cross the reference plane definition at C. Because of the variety of board and mounting schemes available, a free-space model of the DUT has limited value.

The third way is to view both the leads and microstrip lines underneath them as part of the DUT. Taking in both the leads and microstrip

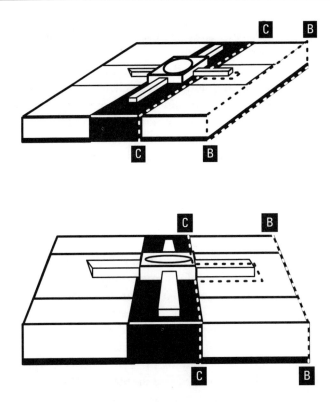

Figure 5.9 Defining the reference plane around a leaded DUT. The C reference plane includes the DUT and its leads but not the microstrip lines underneath.

lines (C′-C and C-C′ planes) is more realistic since the DUT must always be mounted to an RF line in order to be used. To partition the reference planes this way, add an extension line of length d to the fixture's midsection (C′-C′) during calibration. Another way to do this is to add d line lengths to all the calibration standards. Following calibration, the physical and electrical reference planes will align at the C′ planes. To study transitions, this same approach can be applied to de-embedding the coaxial connector pins overlapping the microstrip lines at the B planes.

5.4 Two-Tier Calibration

To de-embed the RF test system from the DUT measurement, the best place to perform a calibration is at the DUT-to-test fixture interface (such as the C planes in Figure 5.1). A calibration at this point will de-embed both the RF test system and the test fixture. The difficulty is that calibration standards for the test fixture are difficult to build. For the best calibration accuracy, the calibration standards should fit in the fixture the same as the DUT. The problem is finding the VNA calibration coefficients for in-fixture calibration standards. Since the standards are uniquely designed for the test fixture, measuring their parasitics is not easy. On the other hand, high-quality coaxial standards are available and can be used to calibrate the test system up to the connection to the test fixture (i.e., the A planes in Figure 5.1).

When well-characterized, in-fixture calibration standards are not available, consider a two-tier calibration. First perform a calibration to the ends of the coaxial cables connecting the RF test system to the test fixture (the A planes). Then perform a second calibration at the DUT interface (the C planes). The two calibration tiers mathematically cascade as error adapters (see Section 2.3.2 in Chapter 2). Using a traditional coaxial calibration kit to calibrate to the ends of the coaxial cables eliminates gross RF errors due to the test system. The availability of coaxial standards makes this step easy. A second calibration at the DUT-to-test fixture interface de-embeds the test fixture's contribution from the measurement. Performing a two-tier calibration minimizes shortcomings due to poor in-fixture calibration standards. A number of methods focus on calibrating the second tier [15, 16].

When the DUT is wire-bonded into the test fixture, so must the calibration standards be during calibration. Making and breaking the bond wires to swap standards increase repeatability error. The varying length of a bond wire adds uncertainty to the calibration. Connecting and reconnecting standards at the first-tier coaxial interface can add as much error as replacing wire-bonded standards at the second tier, compounding repeatability problems. Rather than inserting multiple standards, one solution is to wire-bond a single active standard such as a diode in place of the DUT [17]. By adjusting the bias, the active standard presents different impedances to the test fixture (see Section 8.6.5 in Chapter 8). A varactor diode presents a precise capacitance at a known current, while the PIN diode can approximate either an open, short, or a resistance, depending on the current. This has good repeatability. The bonding scheme of the diode should emulate that of the DUT.

5.5 Test Fixture Calibration

5.5.1 Calibration Standards: Coaxial Versus In-Fixture

As discussed in Section 5.4, calibrating the test fixture to the DUT's pads requires in-fixture calibration standards. To appreciate the difficulties, a comparison of how in-fixture standards are physically and electrically different from coaxial standards is useful.

Coaxial standards are surrounded by ground, serving to shield the center pin. Conversely, in-fixture standards are usually on a substrate mounted in place of the DUT. Substrate elements are electrically small, which means they can be approximated as lumped elements, simplifying their characterization. At high frequency, arriving at their parasitic values is simpler than with larger coaxial standards. Substrate resistors tend to be thin-film and less than a skin-depth thick. This makes their high-frequency resistance more predictable than with high-frequency coaxial resistors. The S-parameters of lumped elements tend to vary smoothly with frequency. When a short approaches a wavelength long, it radiates. In-fixture standards can couple to nearby metal in the fixture. Coupling and radiation are nonsystematic errors that are difficult to de-embed.

Developing in-fixture standards can be the most difficult part of building a test fixture. They are custom-designed and usually require many iterations to build precisely. When designing calibration standards, the properties to focus on are attenuation, characteristic impedance, and the parasitics (inductances and capacitances). Calibration standards should be stable over temperature and mechanically repeatable. Most importantly, their response should not change each time they are inserted.

Some general guidelines can ease the design of in-fixture standards. Alumina, sapphire, silicon, or GaAs are common substrate materials to choose from. The material's dielectric will impact the behavior of the standard [i.e., the delay of the thru (see Section 2.4.5 in Chapter 2)]. At frequencies greater than 2 GHz, silicon substrates exhibit higher loss and parasitic capacitance than other materials (see Section 6.1).

Employing a photolithographic, thin-film fabrication process yields well-defined standards. When precision is not critical, a thick-film fabrication is less expensive. After fabrication, there must be a way to verify the standards other than the test fixture. When the standards are SOLT, try a TRL-calibrated fixture to independently characterize them. A good load will have constant magnitude and a linear, broadband phase response. When creating 50-Ω CPW load standards, always choose two 100-Ω resistors in parallel over one 50-Ω. This approach cuts the parasitic pad inductance in half.

5.5.2 Calibration Methods

The calibration methods listed in Section 2.6 in Chapter 2 (i.e., SOLT, TRL, and LRM) are also available to calibrate RF test fixtures. Either the physical design of the fixture or the ability to construct in-fixture standards determines the method. In general, which method to select will depend on the flexibility of the test fixture's design, particularly the style of its internal and external ports.

As discussed in Section 2.6.1, the quality of SOLT standards ultimately limits the quality of the SOLT calibration achievable. If the SOLT standards cannot be independently characterized, then their RF behavior may be found at the same time as the test fixture's behavior (see Section 5.1.2). To do so, begin by sketching equivalent circuits of the test fixture and standards. Insert the standards into the fixture one at a time and measure. Next, use software to optimize the equivalent circuits of each standard in the fixture until they agree with the measured data. The equivalent-circuit models should faithfully represent the fixture and standards, otherwise the optimization routine will elicit confusing results. Making assumptions about the fixture, such as its symmetry (that is, the ports are reciprocal), can speed the optimization process.

When the standards are characterized at the same time as the fixture, defining the reference plane can be a source of confusion. The dilemma is knowing where to separate the RF behavior of the standards from the fixture. Using at least one well-defined standard helps locate the fixture-to-standard reference plane.

At millimeter waves, TRL lines become short, demanding tight fabrication tolerances. In such cases, TRA can be a better choice [18]. Computationally similar to TRL, TRA relies on a precision attenuator instead of a line to define the characteristic impedance Z_0. Attenuators are easier to fabricate at millimeter-wave frequencies than lines of a set electrical length. To build an attenuator, lay out a line resembling a spiral inductor (see Figure 2.13 in Chapter 2). Depositing thin-film resist in the areas in between the conducting coils will decrease the edge coupling between spirals. In fact, surrounding all the standards with thin-film resist in the unused areas of the substrate dampens undesired substrate modes.

There is another calibration method resembling TRL that is well suited to test fixtures that have immovable launches [19]. A problem with TRL is that the line is longer than the thru standard. To accommodate the line requires moving the fixture's ports further apart. One solution is to replace the line with two thrus, each having a discontinuity (see Figure 5.10). By placing the same discontinuity at two different locations, the line standard can be eliminated. All the calibration standards are then the same length. Like TRL, the length l should be a quarter-wavelength ($\lambda/4$) at the center frequency.

5.5.3 Calibration Phase Uncertainty

Phase uncertainty in the measurement continues to exist even after calibration [20]. For instance, when a calibration method assumes symmetry in the input and output ports, it uses the following equation during computation

$$S_{12} = S_{21} = \pm\sqrt{S_{12}S_{21}} \tag{5.1}$$

In (5.1), the sign can be chosen either way, leading to ±90° phase uncertainty. In some measurements, the amount of phase shift introduced by the test fixture may need to be determined.

To decide which sign to use, more information is required than is ordinarily gained during calibration. To discover the phase shift, perform a second calibration using a calibration method different from the first. The sole purpose of the second calibration is to determine the phase at the DUT's reference plane inside the fixture. Review the calibration equations of the second method to specifically solve them for the phase. Sometimes the initial calibration method is overdetermined (i.e., has more information

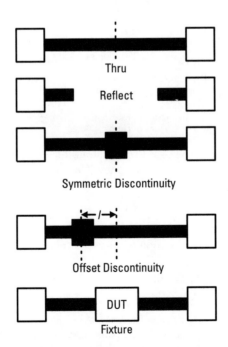

Figure 5.10 A calibration method useful for test fixtures with fixed launchers. The discontinuities are shifted one from another by a length *l*.

than needed to solve the calibration equations). In this instance, the additional information may be used to find the phase at the calibration plane.

5.6 Summary

Test fixtures are often perceived as a headache, usually built as an afterthought in the design process. In order for designers to deliver high-performance RF products, the products must be qualified using a reliable test fixture. This chapter explains how to design RF test fixtures. The discussion has been general since the type, size, and number of RF components to test are nearly limitless. RF products' size, weight, and cost are driving the need for smaller and more integrated packaged components. Shorter product design cycles mean that the time to design and build test fixtures is getting shorter. These trends will not make the job of the test fixture designer any easier.

References

[1]   Partha, R., "Use Matrix Models to Make Analysis Easy for Microstrip Matching Circuits," *RF Design*, Vol. 20, No. 9, 1996, pp. 50–62.

[2]   Bauer, R., and P. Penfield, "De-embedding and Unterminating," *IEEE Trans. on Microwave Theory and Techniques*, Vol. 22, No. 3, 1974, pp. 282–288.

[3]   Agilent Technologies, "In-Fixture Measurements Using Vector Network Analyzers," *Agilent Application Note AN 1287-9*, 1999.

[4]   Williams, D., "De-embedding and Unterminating Microwave Fixtures with Nonlinear Least Squares," *IEEE Trans. on Microwave Theory and Techniques*, Vol. 38, No. 6, 1990, pp. 787–791.

[5]   Ross, P., and B. Geller, "A Broadband Microwave Test Fixture," *Microwave Journal*, Vol. 30, No. 5, 1987, pp. 233–248.

[6]   Eisenhart, R.L., "A Better Microstrip Connector," *IEEE Microwave Symposium Digest*, 1978, pp. 318–320.

[7]   Krems, T., et al., "Millimeter-Wave Performance of Chip Interconnections Using Wire Bonding and Flip Chip,*" IEEE Microwave Symposium Digest*, 1996, pp. 247–250.

[8]   Dalman, G., "New Waveguide-to-Coplanar Waveguide Transition for Centimetre and Millimetre Wave Applications," *Electronics Letters*, Vol. 26, No. 13, 1990, pp. 830–831.

[9]   Ponchak, G., and R. Simons, "A New Rectangular Waveguide to Coplanar Waveguide Transition," *IEEE Microwave Symposium Digest*, 1990, pp. 491–492.

[10] Godshalk, E., "A V-Band Wafer Probe Using Ridge-Trough Waveguide," *IEEE Trans. on Microwave Theory and Techniques*, Vol. 39, No. 12, 1991, pp. 2218–2228.

[11] Romanofsky, R., and K. Shalkhauser, "Fixture Provides Accurate Device Characterization," *Microwaves & RF*, Vol. 30, No. 3, 1991, pp. 139–148.

[12] Izadian, J., "Unified Design Plans Aid Waveguide Transitions," *Microwaves & RF*, Vol. 26, No. 5, 1987, pp. 213–222.

[13] Liu, S., and G. Boll, "Probe Design Extends On-Wafer Testing to 120 GHz," *Microwaves & RF*, Vol. 32, No. 6, 1993, pp. 104–110.

[14] Prabhu, A., and N. Erickson, "A 40 to 50 GHz HEMT Test Fixture," *IEEE Microwave Symposium Digest*, 1995, pp. 427–429.

[15] Vaitkus, R., "Wideband De-embedding with a Short, an Open, and a Through Line," *Proc. IEEE*, Vol. 74, No.1, 1986, pp. 71–74.

[16] Silvonen, K., "Calibration of Test Fixtures Using at Least Two Standards," *IEEE Trans. on Microwave Theory and Techniques*, Vol. 39, No. 4, 1991, pp. 624–630.

[17] Lane, R., "De-embedding Device Scattering Parameters," *Microwave Journal*, Vol. 27, No. 8, 1984, pp. 149–156.

[18] Soares, R., and P. Gouzien, "A Novel Very Wideband 2-Port S-Parameter Calibration Technique," *Microwave and Optical Technology Letters*, Vol. 3, No. 6, 1990, pp. 210–212.

[19] Wan, C., et al., "A New Technique for In-Fixture Calibration Using Standards of Constant Length," *IEEE Trans. on Microwave Theory and Techniques*, Vol. 46, No. 9, 1998, pp. 1318–1320.

[20] Zhu, N., "Phase Uncertainty in Calibrating Microwave Test Fixtures," *IEEE Trans. on Microwave Theory and Techniques*, Vol. 47, No. 10, 1999, pp. 1917–1922.

6

On-Wafer Characterization

Before the advent of coplanar probes, finding the RF behavior of a die on a wafer was a complicated process. First the wafer was singulated and an individual die mounted into a test fixture. Only then could the die's (and, more generally, the wafer's) RF performance be known. Fixturing involved attaching the die to a PCB, wirebonding to the bond pads, connecting RF cables to the fixture, and measuring. Discriminating between the die's and the fixture's responses became a central issue. Furthermore, fixturing die was a time-consuming process, making it impractical for high-volume screening. Thus arose the need for on-wafer characterization.

What is the best way to RF-probe a die on a wafer? The RF characterization techniques explained in this chapter answer this question. Some on-wafer devices are active (such as transistors and diodes) while others are passive (such as resistors, inductors, and capacitors). Electrical models of the devices are built using RF measurements made on-wafer. Circuits are designed with these models, so it is important that the on-wafer characterization be accurate. Calibration, discussed in Chapter 2, is just the first step. De-embedding must also be performed to discover the RF performance of the on-wafer DUT. Before examining the de-embedding process, the electrical behavior of a wafer substrate should first be discussed.

6.1 Conductive Versus Insulating Substrates

A brief discussion of wafer types and their RF properties helps understand why de-embedding the wafer's parasitics is so essential. Most substrates are made of either silicon (Si) or GaAs. These substrates can be either semi-insulating or semiconductive, depending on the doping. Lightly doped silicon substrates have resistances around 10 KΩ-cm. This is significantly below what is achievable with lightly doped GaAs (10 MΩ-cm).

Understanding the wafer's loss mechanisms helps to understand the RF behavior of a die on the wafer. In bulk semiconductors, there are two types of losses: ohmic loss and polarization loss. The loss associated with carriers traveling through a semiconductor is ohmic loss. The dielectric polarization of the substrate has a frequency dependence, leading to a frequency-dependent polarization loss. At low frequencies, ohmic loss dominates, while at high frequencies, polarization loss does.

These losses, along with skin effect, impact the RF performance of the conductors on the wafer. Skin effect sets up a current return path just below the surface of the substrate (see Figure 6.1). The degree of return (or image) current flowing along the wafer surface depends on the conductivity of the substrate [1]. Interestingly, more ground return current can flow along the substrate's surface than through the coplanar ground conductors. In such cases, the skin losses in the substrate will be larger than the losses in the ground conductor.

During IC processing, an insulating oxide layer is deposited between the substrate and the metal. When the metal-insulator sandwich contacts the substrate, either an accumulation, depletion, or inversion layer forms on the semiconductor surface, depending on the surface potential. Even when the bulk resistivity of the semiconductor is high, the resistivity on the substrate surface can still be low. With conductive substrates, the electric field is concentrated in the oxide above the substrate's surface, giving rise to a large distributed C_{ox}. Since the substrate itself has as a low resistance, both R_{sub} and C_{sub} within the semiconductor are small.

High-frequency losses are directly related to the field penetration into a semi-insulating substrate. At high frequency, multiple modes of electromagnetic propagation can be launched in the substrate [2]. When the electric field does not penetrate deeply into the substrate, a surface wave mode can arise at higher frequencies (see Section 3.1 in Chapter 3). With semi-insulating substrates, the electric field penetrates deeper into the substrate as the frequency increases, establishing a quasi-TEM mode.

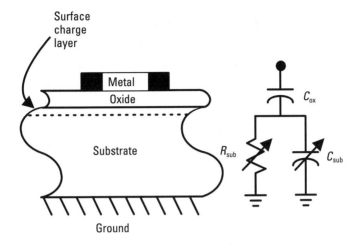

Figure 6.1 A *metal-oxide semiconductor* (MOS) stack. Conductive substrates have low resistance R_{sub} and almost no capacitance C_{sub}, while insulating substrates have a high R_{sub} and a large C_{sub}. The dotted line shows a surface charge layer formed due to a potential difference between the substrate and the metal.

One way to quantify the substrate's loss is with a simple transmission line. With transmission lines, loss can be expressed through the relative dielectric constant ε_r. The insulating oxide layer between the line and substrate lowers ε_r, lessening the line's loss. Because the oxide layer increases the conductor-substrate separation, electromagnetic fields in the substrate will not be as strong [3].

Conductive substrates are lossy and their conductivity is frequency dependent, so characterizing devices can be difficult. On a conductive silicon substrate, wide transmission lines have more high-frequency loss than narrow lines [4]. This characteristic is especially problematic for calibration methods that rely on the characteristic impedance Z_0 of a transmission line (see Section 2.6.3 in Chapter 2) [5]. Conversely, the high resistivity of semi-insulating GaAs leads to less RF transmission line loss than on semi-insulating Si. Whether semi-insulating or semiconducting, a silicon substrate's resistivity increases rapidly with frequency. Hence, its transmission line loss increases in the same manner.

In general, transmission line loss depends on the line width, the substrate's properties, and the thickness of the insulating layer (SiO_2 or polymide) between the metal layer and Si substrate [6]. For example, coplanar waveguide on a conductive silicon substrate with a thick polymide insulating layer can yield minimal transmission line attenuation [7].

6.2 Probe Pads and Interconnecting Lines

IC fabrication engineers are interested in the intrinsic characteristics of the device, such as its doping levels, diffusion depth, and junction depths within the wafer. Capacitance, resistance, and inductance measurements are the best ways to reveal these. Placing the coplanar probes directly on the device to measure it is not possible. Instead, measurement requires probe pads and interconnect lines leading to the DUT. The disadvantage is that the parasitics of the pads and interconnects can be larger than the device measurement itself.

The act of calibration defines the reference plane at the probe tips. To de-embed the probe pads and interconnects and find the intrinsic DUT's RF behavior, the electrical reference plane must be shifted from the coplanar probe pads to the DUT. To illustrate the impact of the pads and interconnects on the device measurement, consider h_{21}, the current gain of a transistor with a shorted output, defined as

$$h_{21} = y_{21} / y_{11} \qquad\qquad (6.1)$$

Any parasitic capacitance will affect the measured admittances y_{21} and y_{11}, so de-embedding the probe pads and interconnects is critical to coming up with an accurate h_{21} measurement. Failure to do so can impact each of the four S-parameters ($S_{11}...S_{22}$) by 1 to 2 dB. This error compounds when the S-parameters are used in complex calculations, making transistor benchmarks such as f_T and f_{max} appear as much as 25% lower than they actually are. The impact of the pads and interconnects becomes larger as the device's dimensions get smaller, particularly with conductive substrates.

Figure 6.2 shows a typical bond pad and interconnect scheme. The bond pads are on top of a low-loss dielectric insulating layer, either oxide or polymide. Also shown is a first-order electrical model of the substrate, good for less than 3 GHz. The value of the high-Q capacitor C_{pad} can range from 0.1 to 0.3 pF, depending on the dielectric material and the substrate's thickness [8]. The resistor R_{pad} depends on the substrate's resistivity. In the bottom of Figure 6.2, the values of the resistors depend on the amount of resistive coupling through the substrate, while the capacitors depend on the fringing fields between the pads and interconnects.

6.3 De-embedding the Pads and Interconnects

Preparing to characterize a DUT on a wafer is a two-step process. First, a VNA calibration is performed, setting the reference plane at the probe tips.

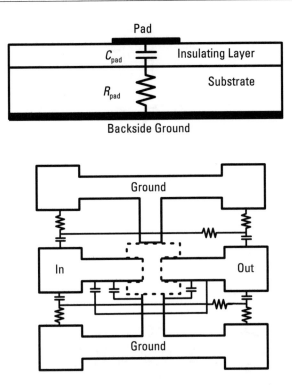

Figure 6.2 Probe pads and interconnecting lines. Inside the dotted line is the DUT. The top figure shows a probe pad on a substrate and its equivalent circuit. The bottom figure shows the probe pad and interconnects along with a schematic of its parasitic circuit elements.

Then, a set of dummy test structures is measured. As will be discussed in this section, these are used to move the electrical reference plane from the probe tips to the DUT plane.

This section outlines five ways, in order of complexity, to de-embed the probe pads and interconnects (see Figures 6.3 and 6.4) [9]. They can be applied to passive structures (inductors, capacitors, and resistors), as well as active devices (transistors and diodes) on the wafer. In general, the de-embedding patterns should have the same metal pattern as the DUT layout. Because substrate resistivity and oxide thickness vary across the wafer, the dummy patterns should be located near the DUT.

6.3.1 Open

Either of the open structures shown in Figure 6.3(a) can be used to de-embed the pads' and interconnects' shunt capacitances [for the equivalent circuit,

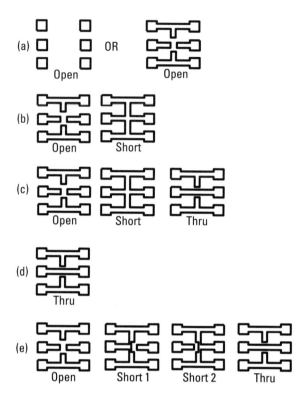

Figure 6.3 The five sets of dummy structures used to de-embed the probe pads and interconnects are (a) open; (b) open and short; (c) open, short, and thru; (d) thru only, and (e) two shorts, open, and thru.

see Figure 6.4(a)]. These capacitances represent the coupling through the substrate [10]. De-embedding solely with an open is valid when the length of the interconnecting lines is short with respect to a wavelength λ.

As mentioned in Section 6.2, pad capacitance is formed by an oxide layer between the metal and substrate (see Figure 6.1). Its value is a function of the oxide thickness and the size of the pad, while the resistance value depends on the resistivity of the semiconductor.

To de-embed the probe pads only, use the open structure on the left in Figure 6.3(a). The probe pads at port 1 (P1) and port 2 (P2) can be described electrically as parallel admittances $y_{p,1}$ and $y_{p,2}$, respectively. Specifically, each admittance can be represented as a series resistor-capacitor combination to the ground (see top of Figure 6.2). To de-embed, simply subtract the admittances $y_{p,1}$ and $y_{p,2}$ of the open structure from the measured input and output admittances (y_{11} and y_{22}, respectively). The pads-only open is applicable at

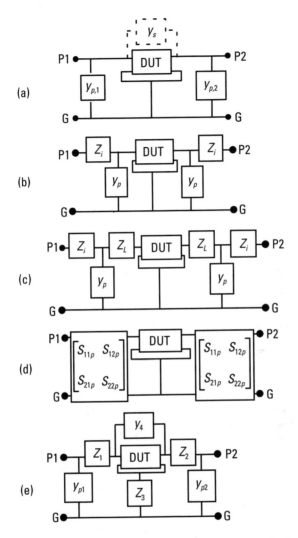

Figure 6.4 The equivalent circuits found with the dummy structures in Figure 6.3 corresponding to (a) open; (b) open and short; (c) open, short, and thru; (d) thru only, and (e) two shorts, open, and thru.

lower frequencies (less than 3 GHz) where the series resistance is small due to a lack of skin effect.

For greater than 3 GHz, use the open structure on the right of Figure 6.3(a) to de-embed the shunt capacitance due to both the probe pads and the interconnects. In Figure 6.4(a), the series admittance y_s (shown in dotted lines) is the contribution of the interconnects. This leads to a π-circuit whose elements are found by inserting the y-parameters of the open $y_{11,open}$, ..., $y_{22,open}$ into [11]

$$y_s = -y_{12,open} \tag{6.2}$$

$$y_{p,1} = y_{11,open} + y_{12,open} \tag{6.3}$$

$$y_{p,2} = y_{12,open} - y_{22,open} \tag{6.4}$$

6.3.2 Open and Short

Adding a short to the previous method de-embeds the series parasitics such as the contact resistance between the probe pad and probe tip, as well as the interconnect's series loss [see Figure 6.3(b)].

To find the circuit elements in Figure 6.4(b), begin the de-embedding process from the outside ports and move in toward the DUT, layer by layer. First subtract the input and output impedances z_i from the measured input and output impedances (z_{11} and z_{22}, respectively). Then convert the result to admittances and subtract y_p. As in Section 6.3.1, the open de-embeds the shunt parasitics

$$z_i = z_{i,short} \tag{6.5}$$

$$y_p = 1 / \left(z_{i,open} - z_{i,short} \right) \tag{6.6}$$

where $Z_{i,short}$ is the series impedance found with the shorted dummy structure and $Z_{i,open}$ is the impedance due to the open. The reference plane is then shifted to the DUT plane. This method is particularly useful when the calibration substrate has metallization different from the wafer's plating, leading to different conductor losses.

6.3.3 Open, Short, and Thru

Supplementing the open and short dummy structures with a thru [see Figure 6.3(c)] better de-embeds the impedance of the interconnect lines [see Figure 6.4(c)]. De-embedding z_i and y_p in the same manner as in Section 6.3.2 yields the input and output impedances z'_{11} and z'_{22}. These are converted to admittances and inserted in (6.7) to find z_L

$$z_L = \eta_{\text{DUT}} / (2 y'_{11}) \tag{6.7}$$

where y'_{11} is the port 1 admittance after de-embedding z_i and y_p. Port 2 uses y'_{22}. The factor η_{DUT} corrects for the DUT gap, denoted by the dotted lines in the bottom of Figure 6.2.

$$\eta_{\text{DUT}} = (l_{\text{FIX}} - l_{\text{DUT}}) / l_{\text{DUT}} \tag{6.8}$$

If l_{FIX} is the length of the interconnecting lines, then l_{DUT} is the total length of the interconnecting lines, including the thru spanning the DUT. Because (6.8) is a ratio, the lengths l_{FIX} and l_{DUT} can be estimated with respect to one another instead of precisely measuring them.

6.3.4 Two-Port Network with a Thru

Characterizing the parasitics of the pads and interconnects becomes more complicated as the frequency increases. Instead of a detailed equivalent circuit, consider describing the pads and interconnects as a two-port network [see Figure 6.4(d)]. To do this, the only structure required is a thru [see Figure 6.3(d)]. The advantage of this method is that complicated pad and interconnect layouts are no problem. The method can even handle conductive substrates.

Assuming the input and output are symmetrical, use the measured S-parameters of the dummy thru ($s_{11}, \ldots, s_{22}$) to find the S_p-matrix

$$s_{11\,p} = s_{22\,p} = \frac{s_{11} + s_{22}}{2 + s_{21} + s_{12}} \tag{6.9}$$

$$s_{12\,p} = s_{21\,p} = \sqrt{\frac{1}{2}(s_{12} + s_{21})(1 - s_{11\,p}^2)} \tag{6.10}$$

Transforming S_p into T-parameters T_p, matrix math can be used to de-embed the DUT from the measured T-parameters T_{meas}

$$T_{\text{DUT}} = T_{p,\text{in}}^{-1} T_{\text{meas}} T_{p,\text{out}}^{-1} = T_p^{-1} T_{\text{meas}} T_p^{-1} \tag{6.11}$$

When the input and output networks are symmetrical, $T_{p,\text{in}}^{-1} = T_{p,\text{out}}^{-1} = T_p^{-1}$. For nonsymmetrical networks ($T_{p,\text{in}}^{-1} \neq T_{p,\text{out}}^{-1}$), the two T-parameters can be found with a short, open, and load mounted in place of the DUT [12].

Be aware that any method that relies on a thru is susceptible to over-calibration when the thru's length is long, usually due to a large DUT in the gap. A long shorting bar across the DUT gap can also cause overestimation. To compensate, see Section 6.3.5.

6.3.5 Two Shorts, an Open, and a Thru

This method improves on the ones described in Sections 6.3.1 to 6.3.3 by de-embedding the port-to-port parasitics and the ground's inductance. Port-to-port coupling can result from fringing fields in the air, as well as from resistive coupling through the substrate [13]. The three-step procedure is as follows.

Start by measuring the open and the thru dummy structures [see Figure 6.3(e)] [14, 15]. Referring to the equivalent circuits of each (see Figure 6.5), the conductances G_1–G_3 are found by measuring their y-parameters

$$G_1 = y_{11,\text{open}} + y_{12,\text{open}} \tag{6.12}$$

$$G_2 = y_{22,\text{open}} + y_{12,\text{open}} \tag{6.13}$$

$$G_3 = \left(-\frac{1}{y_{12,\text{open}}} + \frac{1}{y_{12,\text{thru}}} \right)^{-1} \tag{6.14}$$

Use G_1 and G_2 to de-embed the shunt admittances from the measurement, resulting in

$$y_{11} = y_{11,m} - G_1 \tag{6.15}$$

$$y_{12} = y_{12,m} \tag{6.16}$$

$$y_{21} = y_{21,m} \tag{6.17}$$

$$y_{22} = y_{22,m} - G_2 \tag{6.18}$$

where m denotes the measured y-parameters.

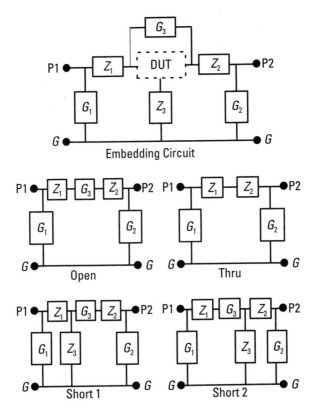

Figure 6.5 Detailed equivalent circuits of each de-embedding structure shown in Figure 6.3(e).

Next, de-embed the series parasitics. Measure the y-parameters of the thru, short1, and short2 dummy structures of Figure 6.3(e) to calculate the impedances

$$Z_1 = \frac{-1/y_{12,\text{thru}} + 1/(y_{11,\text{short1}} - G_1) - 1/(y_{22,\text{short2}} - G_2)}{2} \tag{6.19}$$

$$Z_2 = \frac{-1/y_{12,\text{thru}} - 1/(y_{11,\text{short1}} - G_1) + 1/(y_{22,\text{short2}} - G_2)}{2} \tag{6.20}$$

$$Z_3 = \frac{1/y_{12,\text{thru}} + 1/(y_{11,\text{short1}} - G_1) + 1/(y_{22,\text{short2}} - G_2)}{2} \tag{6.21}$$

Convert the y-parameters in (6.15) to (6.18) to Z-parameters $(Z_{11}, \ldots, Z_{22})$ and subtract Z_1, Z_2, Z_3 using

$$Z_{11,s} = Z_{11} - Z_1 - Z_3 \qquad (6.22)$$

$$Z_{12,s} = Z_{12} - Z_3 \qquad (6.23)$$

$$Z_{21,s} = Z_{21} - Z_3 \qquad (6.24)$$

$$Z_{22,s} = Z_{22} - Z_2 - Z_3 \qquad (6.25)$$

Transform (6.22) to (6.25) into y-parameters to get the DUT's de-embedded y-parameters

$$y_{11}^{DUT} = y_{11,s} - G_3 \qquad (6.26)$$

$$y_{12}^{DUT} = y_{12,s} - G_3 \qquad (6.27)$$

$$y_{21}^{DUT} = y_{21,s} - G_3 \qquad (6.28)$$

$$y_{22}^{DUT} = y_{22,s} - G_3 \qquad (6.29)$$

This method works particularly well with small periphery devices. When the periphery is large, the parasitics of the four dummy structures in Figure 6.3(e) become large. As they add up, overestimation of the device performance can occur [16]. When the substrate is lossy, as with *complementary metal-oxide semiconductor* (CMOS), it helps to supply more detail to the equivalent circuit such as the contact resistance between the probe's metal and the substrate's probe pad [17, 18]. When the DUT layout is symmetrical, it is acceptable to omit one of the dummy short structures.

6.3.6   Some Points to Consider When De-embedding

The following are some points to consider when laying out probe pads and interconnecting lines. The probe pads should be as small as possible to reduce capacitance to the ground and nearby interconnects. To isolate the interconnecting lines from the underlying substrate, the wafer oxide should be as thick as possible (for instance, a field oxide). The substrate should either be highly conductive (to reduce resistance to the backside ground) or highly insulating (to behave principally as a dielectric). The choice of conductivity depends on the circuit application and layout.

The ground traces on the surface should have low inductance [19]. Reserve the DUT area as small as possible, no bigger than the biggest DUT to be characterized. The thru structure used in de-embedding will then be no longer than necessary.

Because the open, short, and thru dummy structures have no DUT, the parasitics of the substrate are somewhat different than with an active device implanted. For best results, the substrate used for the de-embedding structures should be the same as the DUT. Specifically, the doping levels, the epi layer, and the overall wafer thickness should be the same as the DUT so that the substrate's parasitics are properly de-embedded. Designing the pads and interconnects for high isolation can help overcome substrate differences between the DUT and the de-embedding structures (see Section 6.5).

Electrically, the ideal probe pad exhibits small capacitance to the ground and a low conductor resistance, made possible with a small pad size and a thick oxide layer. To further decrease the substrate capacitance, insulating trenches cut deep into the epi and filled with field oxide can be placed directly underneath the probe pads.

What ultimately determines the pad size is the probe, not the wafer fabrication process. Probe overtravel on the pad, as well as the width of the probe tip, decide the minimum pad allowed. Employing a 50-μm probe overtravel is common. With membrane probes, even smaller pads are possible since skating is not a factor (see Section 4.5).

The contact resistance changes every time the probe contacts the pad [20]. While the amount of change is small, it can make an impact when de-embedding minute quantities. Specifically, the equivalent circuit values of the pads and interconnects should not be less than the repeatability of the probe contact.

To find the equivalent circuit of the pad and interconnects more easily, design a layout where the major parasitics are easily identifiable and dominate. Readily identifiable parasitics make the modeling task easier while large, dominating elements make it more accurate. At frequencies below a few gigahertz, the series impedance of the interconnects is low. In such cases, the Z terms in Figure 6.4 can be ignored. Capacitance is small when the interconnects and pads are not close together or when the substrate is semiconductive.

Sometimes, calibration is performed with open pads instead of lifting the probes up in the air. When verifying with probes in the air, an open-pad calibration will appear to have gain. This is due to a lack of inductance in the air compared with the open pads.

Care should be taken when attempting time-domain techniques to shift the reference plane from the probe pads to the DUT. Probe tip placement on the pads inevitably varies, and with such small distances the electrical reference plane can easily be shifted into the DUT.

6.3.7 Should the Transistor Finger Metal Be De-embedded?

There are two approaches to defining the DUT. One is to de-embed all the conductors on the top surface of the wafer (pad, interconnect, and DUT metal), defining the reference plane horizontally between the metal and oxide (see Figure 6.1). This helps to understand what is going on below the wafer surface, where IC processing has the most impact. This type of de-embedding is helpful in device model development where RF quantities can be miniscule. Yet the device cannot be run without metal, so de-embedding this way is not useful for designing circuits. Rather, de-embed up to the edge of the device, cutting the reference plane vertically through the wafer. The DUT then includes some device metal but not the probe pads and interconnecting lines. Because the interconnecting lines are different with each layout, they should be excluded from the DUT model.

Specifically, IC device designers consider the DUT to be the device within the substrate, underneath the metal on the wafer surface. This is the area most affected by epitaxial characteristics such as doping implants, thermal diffusion, and time. To study the intrinsic device, define the area beneath the wafer surface as the DUT. The reference plane will lie horizontally between the metal and oxide layers, excluding the metal on the surface. To define the reference plane, de-embed all the metal on the surface by using a narrow shorting bar [see Figure 6.6(a)]. This will de-embed the metal fingers, including the shorting bar itself. The shorting bar should not be too wide, otherwise it creates inductance and capacitance not present in the actual DUT.

If the desire is to include the metal fingers as part of the DUT, then short them out completely [see Figure 6.6(b)]. The open calibration standard should stop at the edge of the finger array. Making the short standard wide minimizes its resistive and inductive parasitics. In either case, the metal should be thick enough to avoid skin losses and the CPW ground lines should be wide to reduce ground inductance.

A close-up of bipolar base (B) and emitter (E) fingers in Figure 6.7 shows the parasitics to de-embed. Measuring with a narrow shorting bar, as shown in Figure 6.7(a), de-embeds the series parasitic resistances and inductances of the fingers. To find their parasitic capacitances, a corresponding open should include the same finger pattern but without the shorting bar [see Figure 6.7(b)].

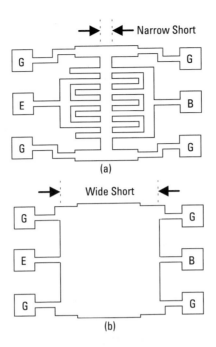

Figure 6.6 Two ways to short a transistor. (a) If the metal fingers are not part of the DUT, then de-embed them with a narrow short. (b) If the fingers are part of the DUT, then use a wide short, setting the reference plane vertically at the edges of the device.

When deciding on whether or not to include the finger metal with the DUT, consider the amount of the quantity being studied. Generally, the more the fingers, the higher the parasitic capacitance (on the order of 1 to 3 fF per finger).

6.3.8 Effect of Pad Parasitics on f_T

Pad parasitics can have a demonstrable effect on the transistor cutoff frequency f_T. In *field effect transistors* (FETs), f_T is defined as

$$f_T = \frac{g_m}{2\pi(C_{gs} + C_{gd})} \cong |h_{21}| - f \qquad (6.30)$$

where g_m is the transconductance, C_{gs} the gate-source capacitance, C_{gd} the gate-drain capacitance, h_{21} the device's gain (a hybrid parameter), and f the frequency. The input admittance measurement y_{11} that captures the gate

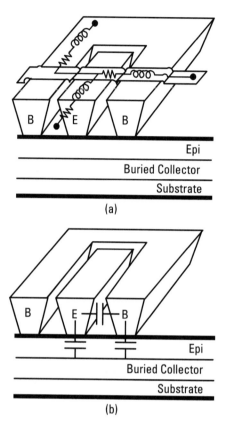

Figure 6.7 Open and shorted transistor fingers. (a) A shorting bar across the fingers and the associated resistances and inductances; and (b) the parasitic capacitances of the open fingers.

behavior is sensitive to parasitics. The parasitic capacitance between the pads and interconnects adds to C_{gs} and C_{gd}. Contact variability of the probes to the pads can also affect the admittances. When defining the reference plane, probe placement error varies from touchdown to touchdown, slightly changing the amount of series inductance L_s (see Section 3.5.3 in Chapter 3). This inductance can resonate with the device's output capacitance C_0, causing $|h_{21}|$ to shift up (see Figure 6.8). C_0 is composed principally of the drain-to-source capacitance C_{DS}. Because C_{DS} is on the order of femptofarads, pad and interconnect parasitics add a pole to the plot around which the upward shift occurs [21]. Making the pads as small as possible is the simplest way to lessen this effect.

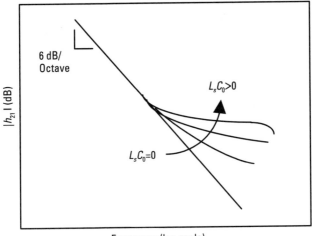

Frequency (log scale)

Figure 6.8 Plot showing how resonance between the series inductance L_s of the probe pads and interconnects and the device's output capacitance C_0 can affect the high-frequency gain h_{21} used to determine f_T.

6.4 De-embedding Pads for Noise

The probe pads and interconnects can strongly affect the noise figure measurement of an on-wafer device. For example, when the input impedance Z_{in} of an FET is large (such as with small-periphery devices), the probe pad becomes a bigger part of the input noise [22]. As the pad grows or the device shrinks, the pad's impedance eventually dominates the noise performance of the FET. This applies equally to semi-insulating GaAs substrates where pad losses are normally small. Because the series parasitics cascade with the DUT's noise performance, the effect of the pad's series parasitics (resistive and inductive) is more important than the shunt parasitics. When trading off the two, it is better to make the series parasitics small compared with the parallel parasitics.

The probe pads are a source of noise, particularly on conductive substrates [23]. As explained in Section 6.1, the pads capacitively couple through the oxide layer to the substrate. The substrate underneath the pad behaves as a resistor, a source of thermal noise. In conductive substrates, normal current flow through the substrate behaves as a noise source, coupling to the pad on the surface. As the pad picks up the thermal noise from the substrate, it degrades the measured noise figure. Putting a grounded metal layer underneath the probe pads will shield the wafer surface from the substrate's noise. By lowering the substrate resistance, a high-Q pad is created whose impedance is primarily capacitive (see Figure 6.2).

Having described the sources of pad noise, a method is needed to split the noise contribution of the pads and interconnects from the FET. To de-embed the noise of the pads and interconnects, view the FET as a noiseless two-port device with noise sources at its input and output. This allows the problem to be tackled through matrix math [24]. Begin with the equation known as the admittance noise correlation matrix

$$[C_A] = 2kT_0 \begin{bmatrix} R_n & \dfrac{F_{min} - 1}{2} - R_n Y_{opt}^* \\ \dfrac{F_{min} - 1}{2} - R_n Y_{opt} & R_n |Y_{opt}|^2 \end{bmatrix} \tag{6.31}$$

where k is the Boltzmann's constant (1.38 by 10^{-23} J/K), T_0 the ambient temperature (290K), R_n the FET's equivalent noise resistance (Ω), F_{min} the minimum noise figure (dB), and Y_{opt} the optimum source admittance (mhos). These fundamental quantities fully characterize the noise of a DUT.

Using one of the methods described in Section 6.3, find the equivalent circuits for the probe pads and interconnects to the gate and drain. For instance, the ABCD matrices $[A]$ of the gate and drain circuits shown in Figure 6.9 are

$$[A_{gate}] = \begin{bmatrix} -1 & -(j\omega L + R_t) \\ -Y_{open} & -Y_{open} \bullet (j\omega L + R_t) - 1 \end{bmatrix} \tag{6.32}$$

$$[A_{drain}] = \begin{bmatrix} -Y_{open} \bullet (j\omega L + R_t) - 1 & -(j\omega L + R_t) \\ -Y_{open} & -1 \end{bmatrix} \tag{6.33}$$

where

$$Y_{open} = \frac{s^2(C_{pad1}\tau_2 + C_{pad2}\tau_1) + s(C_{pad1} + C_{pad2})}{\tau_1\tau_2 s^2 + (\tau_1 + \tau_2)s + 1} \tag{6.34}$$

and $\tau_1 = C_{pad1} \bullet R_{pad1}$ and $\tau_2 = C_{pad2} \bullet R_{pad2}$. Transform $[A_{gate}]$ and $[A_{drain}]$ to their Y-parameter equivalents $[Y_{gate}]$ and $[Y_{drain}]$. Taking the real part of each $[Y]$ and

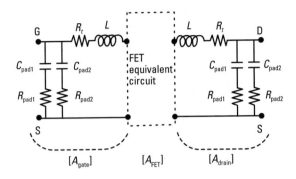

Figure 6.9 The equivalent circuits of the gate (G) and drain (D) probe pads and interconnects. Shown below it are the ABCD matrices [A] for each section, cascaded to get the measured response $[A_{gate}] \cdot [A_{FET}] \cdot [A_{drain}] = [A_{meas}]$.

inserting it into (6.31) gets the admittance correlation matrices $[C_{Y,gate}]$ and $[C_{Y,drain}]$ using

$$[C_Y] = 2kT \cdot \mathrm{Re}[Y] \qquad (6.35)$$

The next step is to convert the correlation matrices $[C_{Y,gate}]$ and $[C_{Y,drain}]$ into ABCD noise matrices. Multiply $[C_{Y,gate}]$ and $[C_{Y,drain}]$ by the transformation matrix $[T]$ to convert the correlation matrices $[C_{A,gate}]$ and $[C_{A,drain}]$ from admittance form to ABCD form

$$[C_A] = [T][C_Y][T]^+ \qquad (6.36)$$

where the transformation matrix $[T]$ and its transpose $[T]^+$ are

$$[T] = \begin{bmatrix} 0 & a_{12} \\ 1 & a_{22} \end{bmatrix} \qquad (6.37)$$

$$[T]^+ = \begin{bmatrix} 0 & 1 \\ a_{12} & a_{22} \end{bmatrix} \qquad (6.38)$$

a_{12} and a_{22} are elements of the two-port ABCD (or chain) matrix, defined as

$$[A_{meas}] = \begin{bmatrix} a_{11} & a_{12} \\ a_{21} & a_{22} \end{bmatrix} \qquad (6.39)$$

The noise correlation matrix for the intrinsic FET $[C_{A,\text{FET}}]$ is then found by

$$[C_{A,\text{FET}}] = [A_{\text{gate}}]^{-1}([C_{A,\text{tot}}] - [C_{A,\text{gate}}])[A_{\text{gate}}]^{+^{-1}} - A_{\text{FET}} C_{A,\text{drain}} A_{\text{FET}}^{+} \quad (6.40)$$

where all the terms on the right side of (6.40) are known. The matrix $[C_{A,\text{tot}}]$ in (6.40) is the measured noise performance of the FET embedded in pads and interconnects. Find R_n, $F_{\min}$, and Y_{opt} of the FET by equating $[C_{A,\text{FET}}]$ found in (6.40) to (6.31).

Another noise de-embedding method outlined by Chen and Deen [25] does not rely on knowing the equivalent-circuit models $[A_{\text{gate}}]$ and $[A_{\text{drain}}]$. Rather, it uses the measurement of an open and two thrus to de-embed the noise reference plane from the probe tips to the FET. When characterizing packaged die, the package itself can generate thermal noise [26]. Proceed in a manner similar to that used with pads and interconnects, de-embedding the package's noise contribution to find the noise performance of the packaged die.

6.5 Measuring High-Isolation Devices

As discussed in the previous sections, RF signals traveling along the probe pads will couple into a low-resistivity substrate (for instance, a silicon biC-MOS substrate). When this occurs, the RF signals will conduct through the substrate from input to output, circumventing the circuit on top. Such substrate leakage is usually greater than leakage by fringing fields between coplanar probes. In general, substrate coupling reduces the DUT's gain, increases its noise figure, and degrades port-to-port isolation.

A grounded metal "shield" surrounding the DUT helps reduce probe pad coupling to the substrate (see Figure 6.10). The bottom metal layer contains the ground shield. Widening the ground shield area lessens its inductance in the ground path. Vias connect the coplanar ground pads through the oxide to the shield below. Connecting the ground pads to the shield ensures a consistent ground reference to all ports of the circuit. Figure 6.11 shows two ways to create a shield around the DUT. Using all available metal layers under the pads yields better pad-to-pad planarity when coplanar probing.

When a grounded metal shield is not achievable, encircle the DUT with a p^+ implant "moat" in the substrate [see Figure 6.11(b)]. The moat connects to the ground probe pads on the top surface [27]. Electrically, the moat acts as a small shunt resistor. This resistor is in parallel with the pad-

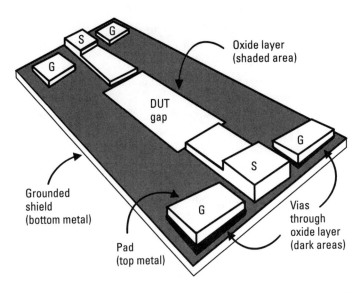

Figure 6.10 A grounded metal shield surrounding the DUT improves isolation. The probe pads are on the top metal. An oxide layer (shaded area) exists between the top and bottom metal layers. Vias (dark areas) connect the coplanar ground pads through the oxide to the ground shield.

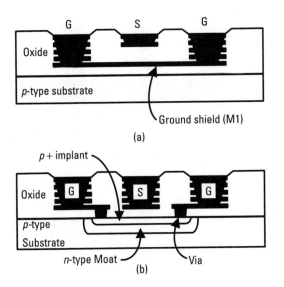

Figure 6.11 Two methods of shielding the DUT. (a) An M1 metal shield surrounding the DUT similar to Figure 6.10; and (b) a doped moat in the substrate also surrounding the DUT. Both inhibit coupling through the substrate.

to-substrate capacitance, effectively shorting it out [28]. A *p*-type implant is conductive, while an *n*-type well will isolate the DUT from the rest of the substrate [29]. Of the two approaches shown in Figure 6.11, Figure 6.11(a) is preferable. It has less parasitic capacitance and less resistance in the ground path compared with a moat.

A third approach is to completely encircle the DUT with grounded vias, building a virtual Faraday cage around it [30]. The ring of vias cuts completely through the substrate to the ground plane on the wafer backside. On the topside, a metal ring around the DUT connects the vias together.

Regardless of the DUT's size (or periphery), one set of shielded dummy structures is enough to de-embed all sizes of devices on the wafer [31]. Otherwise, repeating the same de-embedding structures for each size DUT will take up valuable space on the wafer. Referring to Figure 6.10, size the DUT gap to fit the largest DUT on the wafer. Keeping the pads and interconnect routing the same, scale the gap to fit the DUT. Using this approach, the parasitics of the embedding structure become well defined and fixed. Because they do not scale with the DUT, the same parasitics can be de-embedded for all size DUTs.

6.6 Characterizing Vertical Devices

Vertical devices are those with one RF contact on the top and the other on the bottom of the substrate. For example, a bipolar transistor can be built either vertically or horizontally (see Figure 6.12). When the collector runs through a buried layer, a horizontal device is formed whose RF connections are all on the top surface of the wafer. On the other hand, power transistors are often made with the collector on the back of the substrate to reduce the collector resistance. The difficulty is in discovering how to use a RF coplanar probe to contact both the top and bottom surfaces at the same time.

One solution is to access an adjacent device. To illustrate the concept, consider two vertical diodes [32, 33]. Figure 6.13 shows an S-G coplanar probe making contact to two diodes on a wafer. The ground probe contacts one diode and the signal probe contacts the DUT. The diode probe pads align with the S-G probe pitch.

To bias the ground and signal probes independently, each is dc-isolated using two capacitors. One capacitor is inserted in the dc path of the signal line, inhibiting current from flowing back toward the RF test system. Another capacitor is inserted to dc-block the ground probe from the coaxial cable's shield. This double-blocking arrangement permits independent biasing of

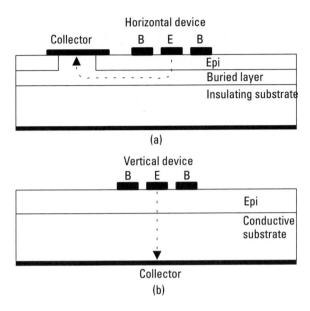

Figure 6.12 (a) Horizontal and (b) vertical bipolar transistors. The dotted line shows current flow.

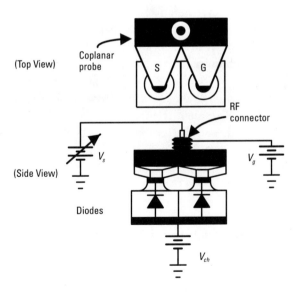

Figure 6.13 Using a coplanar probe to RF characterize a vertical diode. The diode under the signal probe tip is the DUT, while the one under the ground probe serves as a path to the wafer's backside ground.

the signal (V_s) and ground (V_g) probes. Applying a chuck potential (V_{ch}) keeps leakage from the chuck from flowing to the wafer's backside.

By fully biasing the diode beneath the ground probe, it presents a small resistance through the wafer to the backside. De-embedding the ground diode from the RF measurement yields just the DUT. Because the diodes are located adjacent on the wafer, one can assume their characteristics are similar. Biasing the two diodes identically, the DUT becomes two diodes in series with nearly the same characteristics.

6.7 Characterizing Passive Components

While active devices have been used as examples, the same techniques apply equally to passive components. This includes not only simple, building-block elements such as resistors, inductors, and capacitors but also passive networks such as filters and attenuators. To illustrate, Figure 6.14 shows a spiral inductor laid out in four different configurations. To connect to the middle of the inductor, IC fabrication processes often have an air bridge to the center of the spiral. To characterize the spiral only, the probe pads should be laid out similar to Figure 6.14(a). A disadvantage when using coplanar probes is that the spiral can magnetically couple to the bottom of a coplanar probe. To avoid this, the layout shown in Figure 6.14(b) is recommended. It includes the contribution of the air bridge.

A general design requirement for inductors and capacitors is that they have a high quality factor Q [34]. Probe pads and interconnects will affect the Q measurement similar to the effect they have on f_T of active devices (see Section 6.3.7). The self-resonant frequency of capacitors and inductors is impacted by probe pads and interconnects. In particular, an inductor's Q will suffer from such parasitics when it is deposited on a low-resistivity substrate [35]. De-embedding using one of the methods outlined in Section 6.3 will uncover the inductor's true Q.

6.8 Millimeter-Wave Characterization

At millimeter-waves, much of the philosophy of RF characterization explained until now remains unchanged. The fundamental difference is that all the elements in the equivalent circuit become distributed rather than lumped. The transition from lumped to distributed occurs when an element's physical length is a significant fraction of a wavelength (such as $\lambda/10$ or $\lambda/4$).

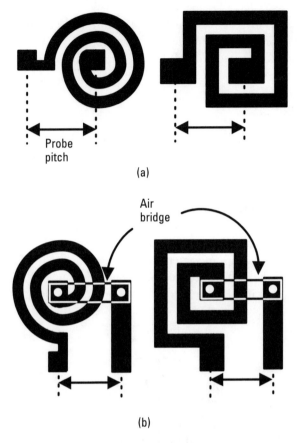

Probe
pitch

(a)

Air
bridge

(b)

Figure 6.14 Coplanar probing spiral inductors. (a) For inductors with no air bridge to the center of the spiral, select a probe pitch to match the spiral. (b) When an air bridge connects to the center of the spiral, add another probe pad.

The precise boundary between lumped and distributed will depend on the design.

When using coplanar probes, the probe tip pitch must narrow as the frequency increases to reduce probe loss. On the wafer, the ground probe pads often have vias connecting to the backside ground metallization. Yet the via's inductance becomes prohibitive at higher frequencies. A solution to using ground vias is to instead use a $\lambda/4$ stub. Looking at Figure 6.15, the round-trip length for a wave traveling through the stub is $\lambda/2$. This causes the reflected wave to cancel with the incident wave, creating a virtual ground. Although narrowband, using a $\lambda/4$ stub has less loss at millimeter waves than other CPW ground pad techniques. Due to skin effect, the stub will have increasing return loss and linear phase shift as the frequency increases. Stub radiation and dispersion are also possible. Shown by dotted lines in Figure 6.15, a full wraparound radial stub permits a broader bandwidth [36].

6.9 Summary

This chapter explains RF on-wafer device characterization techniques to enable more accurate measurements. Theoretically, on-wafer measurements should yield the same results as those made using a test fixture, the functional difference being CPW probes instead of RF launchers and interconnects. While coplanar probes avoid many of the parasitics found in test fixtures, similar issues arise. Coplanar probes cannot contact the DUT directly, so probe pads and interconnects must be de-embedded, not unlike de-embedding the launchers and interconnects in a test fixture. Understanding what goes on at the DUT interface lays the foundation for a broader discussion of test systems in Chapter 7.

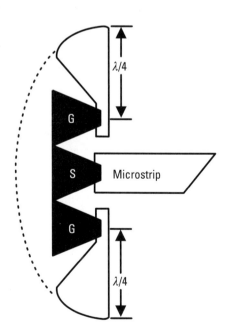

Figure 6.15 $\lambda/4$ stubs used for coplanar ground pads. The dotted line shows a wraparound radial stub used instead of two $\lambda/4$ stubs.

References

[1] Zheng, J., V. Tripathi, and A. Weisshaar, "Characterization and Modeling of Multiple Coupled On-Chip Interconnects on Silicon Substrate," *IEEE Trans. on Microwave Theory and Techniques*, Vol. 49, No. 10, 2001, pp. 1733–1739.

[2] Hasegawa, H., M. Furukawa, and H. Yanai, "Properties of Microstrip Line on Si-SiO2 System," *IEEE Trans. on Microwave Theory and Techniques*, Vol. 19, No. 11, 1971, pp. 869–881.

[3] Reyes, A., et al., "Silicon as a Microwave Substrate," *IEEE Microwave Symposium Digest*, 1994, pp. 1759–1762.

[4] Zaage, S., and E. Groteluschen, "Characterization of the Broadband Transmission Behavior of Interconnections on Silicon Substrates," *IEEE Trans. on Components, Hybrids, and Manufacturing Technology*, Vol. 16, No. 7, 1993, pp. 686–691.

[5] Williams, D., and R. Marks, "Accurate Transmission Line Characterization," *IEEE Microwave and Guided Wave Letters*, Vol. 3, No. 8, 1993, pp. 247–249.

[6] Kleveland, B., T. Lee, and S. Wong, "50-GHz Interconnect Design in Standard Silicon Technology," *IEEE Microwave Symposium Digest*, 1998, pp. 1913–1916.

[7] Ponchak, G., A. Margomenos, and L. Katehi, "Low-Loss CPW on Low-Resistivity Si Substrates with a Micromachined Polymide Interface Layer for RFIC Interconnects," *IEEE Trans. on Microwave Theory and Techniques*, Vol. 49, No. 5, 2001, pp. 866–870.

[8] Camilleri, N., and J. Kirchgessner, "Bonding Pad Models for Silicon VLSI Technologies and Their Effects on the Noise Figure of RF NPNs," *IEEE Microwave and Millimeter-Wave Monolithic Circuits Symposium Digest*, 1994, pp. 225–228.

[9] Kolding, T., "On-Wafer Calibration Techniques for Giga-Hertz CMOS Measurements," *IEEE International Conference on Microelectronic Test Structures*, Vol. 12, 1999, pp. 105–110.

[10] Lovelace, D., "Program De-Embeds Wafer-Probed Data," *Microwaves & RF*, Vol. 32, No. 6, 1993, pp. 136–138.

[11] Hanseler, J., H. Schinagel, and H. Zapf, "Test Structures and Measurement Techniques for the Characterization of the Dynamic Behavior of CMOS Transistors on Wafer in the GHz Range," *IEEE International Conference on Microelectronic Test Structures*, Vol. 5, 1992, pp. 90–93.

[12] Weng, J., "Error Correction in High-Frequency 'On-Wafer' Measurements," *IEEE International Conference on Microelectronic Test Structures*, Vol. 7, 1994, pp. 164–167.

[13] Wijnen, P., H. Claessen, and E. Wolsheimer, "A New Straightforward Calibration and Correction Procedure for 'On-Wafer' High Frequency S-Parameter Measurements," *IEEE Bipolar Circuits and Technology Meeting*, 1987, pp. 70–73.

[14] Cho, H., and D. Burk, "A Three-Step Method for the De-Embedding of High-Frequency S-Parameter Measurements," *IEEE Trans. on Electron Devices*, Vol. 38, No. 6, 1991, pp. 1371–1375.

[15] Vandamme, E., D. Schreurs, and C. Dinther, "Improved Three-Step De-Embedding Method to Accurately Account for the influence of Pad Parasitics in Silicon On-Wafer Test Structures," *IEEE Trans. on Electron Devices*, Vol. 48, No. 4, 2001, pp. 737–742.

[16] Raskin, J., et al., "Accurate SOI MOSFET Characterization at Microwave Frequencies for Device Performance Optimization and Analog Modeling," *IEEE Trans. on Electron Devices*, Vol. 45, No. 5, 1998, pp. 1017–1025.

[17] Kolding, T., "A Four-Step Method for De-Embedding Gigahertz On-Wafer CMOS Measurements," *IEEE Trans. on Electron Devices*, Vol. 47, No. 4, 2000, pp. 734–740.

[18] Costa, D., W. Liu, and J. Harris, "A New Direct Method for Determining the Heterojunction Bipolar Transistor Equivalent Circuit Model," *IEEE Bipolar Circuits and Technology Meeting*, 1990, pp. 118–121.

[19] Kolding, T., "On-Wafer Calibration Techniques for Giga-Hertz CMOS Measurements," *IEEE International Conference on Microelectronic Test Structures*, Vol. 12, 1999, pp. 105–110.

[20] Carhon, G., et al., "Characterising Differences Between Measurement and Calibration Wafer in Probe-Tip Calibrations," *Electronics Letters*, Vol. 35, No. 13, 1999, pp. 1087–1088.

[21] Jones, K. and E. Godshalk, "Waveguide Probe Tackles V-Band On-Wafer Tests," *Microwaves & RF*, Vol. 29, No. 10, 1990, pp. 99-103.

[22] Deen, M., and C. Chen, "The Impact of Noise Parameter De-embedding on the High-Frequency Noise Modeling of MOSFET's," *IEEE International Conference on Microelectronic Test Structures*, Vol. 12, 1999, pp. 34–39.

[23] Camilleri, N., et al., "Bonding Pad Models for Silicon VLSI Technologies and Their Effects on the Noise Figure of RF NPNs," *IEEE Microwave and Millimeter-Wave Monolithic Circuits Symposium Digest*, 1994, pp. 225–228.

[24] Biber, C., et al., "Technology Independent Degradation of Minimum Noise Figure Due to Pad Parasitics," *IEEE Microwave Symposium Digest*, 1998, pp. 145–148.

[25] Chen, C., and M. Deen, "A General Noise and S-Parameter Deembedding Procedure for On-Wafer High-Frequency Noise Measurements of MOSFETs," *IEEE Trans. on Microwave Theory and Techniques*, Vol. 49, No. 5, 2001, pp. 1004–1005.

[26] Pucel, R., et al., "A General Noise De-embedding Procedure for Packaged Two-Port Linear Active Devices," *IEEE Trans. on Microwave Theory and Techniques*, Vol. 40, No. 11, 1992, pp. 2013–2024.

[27] Kolding, T., "Impact of Test-Fixture Forward Coupling on On-Wafer Silicon Device Measurements," *IEEE Trans. on Microwave and Guided Wave Letters*, Vol. 10, No. 2, 2000, pp. 73–74.

[28] Colvin, J., and S. Bhatia, "A Bond-Pad Structure for Reducing Effects of Substrate Resistance on LNA Performance in a Silicon Bipolar Technology," *IEEE Bipolar Circuits and Technology Meeting*, 1998, pp. 109–112.

[29] Biber, C., et al., "Microwave Frequency Measurements and Modeling of MOSFETs on Low Resistivity Silicon Substrates," *IEEE International Conference on Microelectronic Test Structures*, Vol. 9, 1996, pp. 211–215.

[30] Wu, J., et al., "A Faraday Cage Isolation Structure for Substrate Crosstalk Suppression," *IEEE Microwave and Wireless Components Letters*, Vol. 11, No. 10, 2001, pp. 410–412.

[31] Kolding, T., "Shield-Based Microwave On-Wafer Device Measurements," *IEEE Trans. on Microwave Theory and Techniques*, Vol. 49, No. 6, 2001, pp. 1039–1044.

[32] Allen, J., C. Chen, and D. Klemer, "On-Wafer Measurement and Modeling of Millimeter-Wave GaAs Schottky Mixer Diodes," *IEEE Microwave Symposium Digest*, 1992, pp. 743–746.

[33] Wartenberg, S., and C. Mohr, "Probe On-Wafer Diodes," *Microwaves & RF*, Vol. 40, No. 3, 2001, pp. 91–96.

[34] Ashby, K., et al., "High Q Inductors for Wireless Applications in a Complementary Silicon Bipolar Process," *IEEE Bipolar/BiCMOS Circuits and Technology Meeting*, 1994, pp. 179–182.

[35] Power, J., et al., "An Investigation of On-Chip Spiral Inductors on a 0.6m BiCMOS Technology for RF Applications," *IEEE International Conference on Microelectronic Test Structures*, Vol. 12, 1999, pp. 18–23.

[36] Williams, D., and T. Miers, "A Coplanar Probe to Microstrip Transition," *IEEE Trans. on Microwave Theory and Techniques*, Vol. 36, 1988, pp. 1219–1223.

7

RF Test Systems

Chapter 6 narrowly focuses on the coplanar probe-to-die connection. This chapter broadens the discussion to include the rest of the RF test system, expanding beyond the interface to the DUT. The most popular test systems are covered here, along with tips on how to employ them. The chapter begins by describing a noise test system, highlighting its sensitivities. High-power test systems are widely used to design RF power amplifiers. Also, measuring a die's performance over a range of temperatures is another requirement. Both hot and cold (cryogenic) RF test setups are detailed.

7.1   On-Wafer Noise Testing

This section describes a typical RF noise test system. To better understand, elementary concepts in noise are briefly covered first.

7.1.1   Basic Concepts in Noise

The noise parameters used to describe a two-port device are the minimum noise figure F_{min}, the optimum source admittance Y_{opt} (composed of a real part G_{opt} and an imaginary part B_{opt}), and the equivalent noise resistance R_n. These are calculated from three quantities, the source impedance of the DUT, the DUT's S-parameters, and the reflection coefficient at the DUT's output. To find these quantities, the test system has a noise source, a tuner connected to the DUT's input port, and a noise figure meter at the DUT output port (see Figure 7.1).

Combining S-parameters with a noise figure measurement allows accurate determination of the DUT's noise parameters. The test system's accuracy depends on knowing the source impedance presented to the DUT's input, along with the test system's losses. Finding these out can be difficult with wafer probing systems [1, 2]. The tuner can present any number of impedances to the DUT's input port, identifying the source impedance that gives the minimum noise figure [3]. The noise added by the tuner must be de-embedded, otherwise it will add to the DUT's noise figure calculation. Once tuned for minimum noise figure, a thru is inserted in place of the DUT and the VNA is switched in to measure the tuner's impedance. The VNA can also be switched in to measure the DUT's S-parameters. Using the tuner and noise source, find the source impedance that gives the optimum noise figure. To ascertain the impedance state of the tuner, insert a thru in place of the DUT, switch in the VNA, and measure the S-parameters. The electrical reference plane should be de-embedded to the ports of the tuner.

With low-noise devices, the tuner should present a low impedance to the DUT's input port. The problem is that a 50-Ω cable connects the tuner to the DUT, creating a mismatch that limits the impedance presented to the DUT's input port. In such cases, low-loss, low-impedance coplanar probes

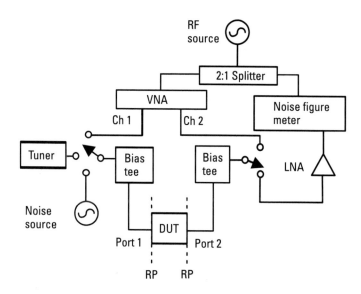

Figure 7.1 A noise measurement setup.

can help (see Section 3.8.2 in Chapter 3). Placing the tuner within a wavelength of the DUT's port delivers the best measurement repeatability. Shortening interconnecting lines minimizes noise generated by RF cables. When short interconnects are not practical, an alternate approach is to construct a coplanar probe with a built-in tuner [4]. The onboard tuner consists of two shunt varactor diodes mounted $\lambda/4$ and $\lambda/2$ from the probe tip. This will give a large variation in impedance as a function of the dc bias applied to the diodes.

Two techniques used for measuring noise are the standard technique and the cold-source technique. In the standard technique, the noise source is alternately switched to its hot (T_h) and cold (T_c) states. At a set power level from the RF source, the noise figure is measured for each state. With the cold source technique, the noise source is switched in during calibration to find out the gain-bandwidth constant of the noise figure meter. In either technique, adjusting the tuner will find the source impedance that gives the lowest noise figure. Keep in mind that component drift in the noise test system means that periodic recalibration will be required [5].

7.1.2 On-Wafer Noise Sources

Anything that has resistance generates noise. The most common on-wafer noise sources are resistors and diodes. To calibrate the noise reference plane to the coplanar probe tips, an on-wafer tantalum nitride (TaN) 50-Ω resistor can be used as a noise source [6]. Hot and cold noise temperatures are generated by running a current through the resistor (creating a hot temperature T_h) or with no current (for a cold temperature T_c, usually room temperature). A reverse-biased diode is also a good on-wafer noise source. PIN diodes can be used for noise sources, as well as Schottky diodes.

An on-wafer diode can also act as a transfer standard [7]. It transfers the noise parameters previously defined at the coaxial connector plane and shifts them to the probe tip plane.

7.1.3 Faraday Shielding

For low-capacitance (less than 100 fF) or low-noise (less than 1-dB NF) measurements, the RF environment around the DUT and wafer must be well contained. The key is to shield the wafer and the chuck from unwanted noise, reducing the stray capacitance to the wafer. Building a Faraday shield around the wafer is a good way to block external fields from impinging on the DUT (see Figure 7.2). An added advantage to using a Faraday-shielded chuck is that it makes the bias pulse's settling time faster. Otherwise, long dc

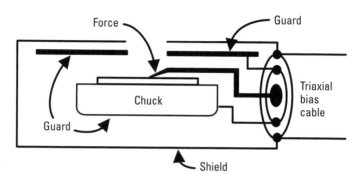

Figure 7.2 Wafer and chuck enclosed in a Faraday shield, assuring an equipotential surface around the wafer and diminishing stray capacitance.

pulses widths can lessen the impact of bias cable capacitance on the RF measurement.

External fields impinge on the wafer from RF radiation through the air or unwanted ground paths near the wafer. Fields capacitively couple to the wafer from nearby grounds, allowing ground noise to sneak onto the wafer. For instance, the power cord of the probe station is plugged into a wall outlet. The plug's ground prong shares ground noise generated within the building. Switching power supplies found inside many commercial electronic products generates noise on the building's ground bus that can upset sensitive RF test equipment. Using an isolation transformer can clean the ground connection from the wall outlet.

The chuck beneath the wafer can couple residual capacitance to the wafer's backside. The amount of capacitance depends on how well the chuck is isolated from the ground. Using a test wafer, the chuck capacitance can be mapped across the chuck and afterwards de-embedded from the final DUT measurement. To map by die location, the test wafer can be an array of capacitors whose values were measured at 1 MHz with an LCR meter.

7.2 High-Power RF Testing

High-power testing is commonly used to find a DUT's output power or to develop a large-signal, nonlinear model of a transistor [8]. During high-power testing, the output power depends on the load impedance connected to the DUT's output port. Sweeping a range of load impedances reveals which value gives the maximum power out (P_{max}) of the DUT. Verifying the accuracy of a large-signal, nonlinear equivalent-circuit model is done using load-pull [9, 10]. A common load-pull setup is shown in Figure 7.3. The variable load attached to the DUT's output is either a mechanical tuner or a solid-state tuner. The DUT's output sees the impedance of the tuner along with mismatches from interconnecting cables and connectors. For on-wafer testing, coplanar probes are located at each RP. An attenuator pad keeps high power from damaging the power meter.

For best results, the losses in the system should be well known and mismatches well characterized to de-embed them from the measured data. Error models for each section's S-parameter performance can be found ahead of time. Cascading the error models gives the DUT's measurement.

To develop a nonlinear model, large-signal S-parameters must be measured on-wafer. A block diagram of a test setup is shown in Figure 7.4. Before beginning, calibrate the RF power at the probe tips using S-parameters instead of a power meter [11]. First, perform a standard one-port calibration (such as SOL) at reference plane 1 (RP_1). Then perform another one-port calibration at RP_2 with a thru in place of the DUT. This gives the S-parameter error model of the path between RP_1 and RP_2. Also measure the power coming out of RP_2. Using the S-parameter error model and knowing the power at RP_2, the electrical reference plane can be mathematically de-embedded to the DUT's ports [12].

The power level of the RF source should adequately drive the DUT while at the same time overcoming the test system's losses [13]. When possible, use a power meter to ensure a known power level at the DUT. Some VNAs have a feature called power flatness that measures the power across a band of frequencies. The RF source is adjusted at each frequency so that the output power at the DUT remains constant over the band.

High current and the accompanying DUT heating can damage coplanar probes. One way to avoid this is to make sure the bias is turned off

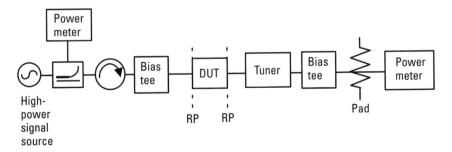

Figure 7.3 Test configuration for performing high-power load-pull measurements.

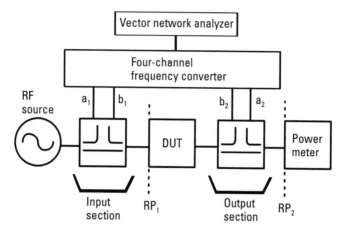

Figure 7.4 Setup for measuring both the S-parameters and the power at the DUT's ports. RP_1 is the reference plane at the DUT input port and RP_2 the reference plane at the power meter. The electrical reference plane at RP_2 is de-embedded back to the DUT's output port.

before lifting the probes, especially when using fast, automated test systems. Accidentally lifting the probes with current flowing can cause a spark to run through the die. Contacting a short-circuited die with bias applied can burn out the coplanar probe tips. Keep in mind the metal composition of both the probe and probe pad. Every metal combination has its own contact resistance, and higher contact resistance values have higher heat dissipation. Also, high-power energy radiated from the top surface of the die will be received by the probes. This can cause feedback and induce oscillations in active devices.

On the probe station chuck, the wafer is held down by a vacuum. Because the wafer is not solidly attached to a heat sink, thermal dissipation on the wafer can be a problem. Die not soldered to a heat sink tend to self-heat rapidly, especially on thick wafers. Pulsing the bias and RF briefly rather than supplying them continuously is one way to keep the DUT from over-heating. In this case, a peak power meter can be used to measure the pulsed RF coming out.

7.3 Characterizing over Temperature

Because operating temperatures on the die can easily range from –65°C to +200°C, RF measurement accuracy at temperature is important. RF measurements over temperature are sometimes extrapolated from RF measurements made at room temperature. A somewhat better approach is to calibrate the test system at room temperature and then make the RF measurements at temperature. The best approach is to both calibrate and measure at temperature.

The coplanar probes, cables, wafer chuck, even the wafer itself expand and contract with temperature. As the chuck expands and contracts, it affects the probe overtravel or skating. Any change in the diameter of the wafer with temperature will impact the spacing between the die, requiring readjustment of the probes at each temperature. The micropositioners that hold the probes will expand and contract, also causing the probes to skate across the pads. There is electrical drift due to probe and cable temperature variations, not to mention test instrument variations.

One way to quantify these changes is to calibrate at the temperature extremes, storing the calibration as CAL_{HighT} and CAL_{LowT}. When measuring the DUT at one of the temperature extremes (either $HighT$ or $LowT$), recall CAL_{HighT} and CAL_{LowT} one at a time without moving the probes. The deviation between the two measurements quantifies the calibration drift over temperature. To ensure the integrity of the measurement, a calibration stability check should be performed hourly during temperature measurements.

In general, recalibrating is good practice for temperature changes greater than 25°C. To measure small RF quantities (such as femtofarads of capacitance), recalibrate for greater than 5°C change. When moving from temperature to temperature, a quick way to check the calibration drift is by measuring an open (such as probes up in the air) at the two temperatures. Looking at a phase-only plot of the return loss (S_{11} or S_{22}), the open's phase difference between the two temperatures equates to $\omega C_0 Z_0$. For example, a temperature difference of 100°C can cause a phase change of 2.4 fF deviation in C_0 of an open in a 50-Ω system.

Regarding the calibration substrate, its dielectric constant will change with temperature, and possibly via-hole integrity through the substrate. The permittivity of an alumina calibration substrate decreases slightly at cryogenic temperatures, affecting the impedance of all the standards it holds.

The calibration standards themselves are affected by temperature. As mentioned in Section 2.4.4 in Chapter 2, the value of the load standard changes over temperature due to the *thermal coefficient* (TC) of the resist. TaN resistors have a TC on the order of $10^{-4}/°C$, so that from 0°C to 125°C, their resistance changes by 1%. This means that all four of the S-parameters will be off by 1% [14, 15]. With each calculation, the 1% error compounds so that the final calibration error can be great [see (2.1) to (2.4) in Chapter 2]. Hence, calibration methods that rely on the quality of the load should be avoided. A better calibration method is TRL, which does not rely on precision standards but rather only on a 50-Ω thru.

7.3.1   Heating the Wafer

The wafer can be heated from underneath using a thermal chuck. During heating, a thermal gradient exists from the chuck through the wafer to its top surface, with the coplanar probes receiving the heat. Figure 7.5 shows a wafer heated to 200°C. When the probe contacts a hot wafer, heat from the wafer surface is transferred through the probe's body. Heat conducts through the probe tip, as well as collecting on the probe's underside, traveling through the body to the coaxial connector. In general, a coplanar probe is more temperature-stable than the RF coaxial cable to which it connects. In fact, the coaxial cable is the most temperature-sensitive part of the RF test system. Temperature changes alter the phase of the RF signal traveling through a coaxial cable. Because of the large thermal gradient in the probe, a new calibration is recommended at each heated temperature. Before calibrating, allow the probes to temperature-stabilize for at least 15 minutes.

At millimeter-wave frequencies, thermal expansion can become a significant fraction of a wavelength, so recalibrating at temperature becomes even more important. The chuck expands upwards as the temperature increases. This impacts the probe overtravel, calling for either larger probe pads or readjusting the probe's overtravel at each temperature.

7.3.2   Cryogenic Cooling

At extremely cold temperatures (below 20K), some transistors exhibit improved RF performance such as higher gain and lower noise figure [16].

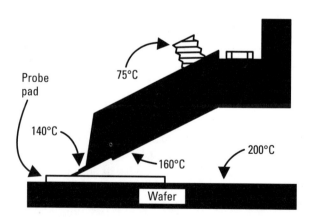

Figure 7.5 Thermal gradient through a coplanar probe. Note the undersurface of the probe body collects more heat than the probe tip contacting the probe pad.

To measure these requires a setup that can calibrate and test in a cold environment.

There are two approaches to cooling the DUT and RF measuring [17]. The most common is putting the test fixture and DUT in a sealed chamber filled with dry nitrogen vapor (see Figure 7.6). This method is useful for temperatures down to 77K. The temperature can be measured either on the chamber walls or with a vapor probe inserted inside. Yet, there still remains an unknown thermal gradient from the nitrogen vapor environment to the DUT itself. At lower temperatures, this gradient becomes significant. Another approach, good down to 10K, is to immerse the entire test fixture, DUT, and cables into a vat of liquid nitrogen.

As the temperature decreases, the ohmic loss of a metal decreases. Contracting metal can cause marginal connections to open. For the test fixture, brass is chosen for its low *thermal coefficient of expansion* (TCE). For the same reason, BeCu cables are particularly stable over temperature [18].

To calibrate the setup in Figure 7.6, the chamber must be opened to swap calibration standards. Closing and recooling the chamber extend the length of time it takes to complete a calibration, and the VNA can drift over time. Also, leaving RF test equipment like the tuner and RF source

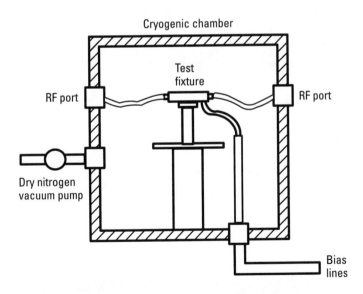

Figure 7.6 Typical cryogenic test setup. The walls of the chamber have cable connector feedthroughs. The VNA is outside the cooling chamber, while the test fixture and DUT are inside.

outside the chamber limits the impedance range that can be presented by the tuner [19].

Following calibration, using offset delay to shift the RP to the DUT is the simplest way to accommodate the variation of the coplanar probes with temperature (see Section 2.5.1 in Chapter 2). First perform a room-temperature calibration. Then measure the return loss of an open at room temperature and at the cryogenic measurement temperature. Looking at phase plots of the two, shift the RP by the phase difference between them. This phase difference accounts for the contraction of the probes and the RF cables during cooling. When temperature-stable RF cables and other components are used, using offset delay to shift the calibration plane from the one found at room temperature is usually deemed acceptable.

To measure the cryogenic temperature, a thermocouple should be attached as close to the DUT as possible. Another way to determine the temperature of the DUT is with an on-wafer diode. At room temperature, apply a dc voltage V and measure the current I_s. Then use the diode relation

$$I_s = I_0 \left(e^{\frac{qV}{nkT}} - 1 \right) \tag{7.1}$$

where q is the electronic charge and k the Boltzmann's constant. Sweeping the voltage, solve (7.1) for the steady-state current I_0 and the ideality factor n. Then, applying the same voltage at the cryogenic temperature, measure I_s and solve for the cryogenic temperature T.

7.4 Summary

This chapter explains how to use three popular RF test systems. Some fundamental aspects cut across all systems, such as calibration. The ideas contained in this chapter, combined with those in Chapters 5 and 6, should allow a high degree of RF measurement accuracy, whether for noise, high power, or over temperature.

References

[1]   Leake, B., A. Davidson, and T. Burcham, "Use Network Data to Hone On-Wafer Noise Parameters," *Microwaves & RF*, Vol. 28, No. 2, 1989, pp. 99–105.

[2]   Pak, S., and T. Chen, "Simple System Yields On-Wafer Noise Parameters," *Microwaves & RF*, Vol. 29, No. 7, 1990, pp. 103–108.

[3] Davidson, A., B. Leake, and E. Strid, "Accuracy Improvements in Microwave Noise Parameter Measurements," *IEEE Trans. on Microwave Theory and Techniques*, Vol. 37, No. 12, 1989, pp. 1973–1977.

[4] Rabjohn, G., and R. Surridge, "Tunable Microwave Wafer Probes," *IEEE GaAs IC Symposium Digest*, 1988, pp. 213–216.

[5] Meierer, R., and C. Tsironis, "An On-Wafer Noise Parameter Measurement Technique with Automatic Receiver Calibration," *Microwave Journal*, Vol. 38, No. 3, 1995, pp. 22–37.

[6] Beland, P., et al., "A Novel On-Wafer Resistive Noise Source," *IEEE Microwave and Guided Wave Letters*, Vol. 9, No. 6, 1999, pp. 227–229.

[7] Dunleavy, L., et al., "Characterization and Applications of On-Wafer Diode Noise Sources," *IEEE Trans. on Microwave Theory and Techniques*, Vol. 46, No. 12, 1998, pp. 2620–2628.

[8] Baureis, P., and D. Seitzer, "Parameter Extraction for HBT's Temperature Dependent Large Signal Equivalent Circuit Model," *IEEE GaAs IC Symposium Digest*, 1993, pp. 263–266.

[9] Kalokitis, D., B. Epstein, and J. Schepps, "Method Permits On-Wafer Load-Pull Testing," *Microwaves & RF*, Vol. 30, No. 10, 1991, pp. 83–92.

[10] Poulin, D., J. Mahon, and J.P. Lanteri, "A High-Power On-Wafer Pulsed Active Load-Pull System," *IEEE Trans. on Microwave Theory and Techniques*, Vol. 40, No. 12, 1992, pp. 2412–2417.

[11] Roth, B., et al., "Vector Corrected On-Wafer Power Measurements of Frequency Converting Two-Ports," *47th ARFTG Conference Digest*, 1996, pp. 14–17.

[12] Ferrero, A., and U. Pisani, "An Improved Calibration Technique for On-Wafer Large-Signal Transistor Characterization," *IEEE Trans. on Instrumentation and Measurement*, Vol. 42, No. 2, 1993, pp. 360–364.

[13] Carbonero, J., et al., "On-Wafer High-Frequency Measurement Improvements," *IEEE International Conference on Microelectronic Test Structures*, Vol. 7, 1994, pp. 168–173.

[14] Anholt, R., *Electrical and Thermal Characterization of MESFETs, HEMTs, and HBTs*. Norwood, MA: Artech House, 1995, p. 49.

[15] D'Almeida, D., and R. Anholt, "Device Characterization with an Integrated On-Wafer Thermal Probing System," *Microwave Journal*, Vol. 36, No. 3, 1993, pp. 94–105.

[16] Meschede, H., et al., "On-Wafer Microwave Measurement Setup for Investigations on HEMTs and High Tc Superconductors at Cryogenic Temperatures Down to 20K," *IEEE Trans. on Microwave Theory and Techniques*, Vol. 40, No. 12, 1992, pp. 2325–2331.

[17] Laskar, J., and M. Feng, "An On-Wafer Cryogenic Microwave Probing System for Advanced Transistor and Superconductor Applications," *Microwave Journal*, Vol. 36, No. 2, 1993, pp. 104–114.

[18] Smuk, J., and J. Wright, "Enhanced Microwave Characterization Technique for Cryogenic Temperatures," *Electronics Letters*, Vol. 20, No. 25, 1990, pp. 2127–2129.

[19] Laskar, J., et al., "Development of Accurate, On-Wafer, Cryogenic Characterization Techniques," *IEEE Trans. on Microwave Theory and Techniques*, Vol. 44, No. 7, 1996, pp. 1178–1182.

8

Package Characterization

RFICs are commonly marketed not as die but as packaged components. Although an essential part of the component, the package is easily overlooked, usually an afterthought to designing the die. A high-performance die in an inferior package results in a mediocre-performing component.

The RF package can be characterized either as empty or with a die inside. Many packages cannot be directly probed, in which case a test fixture provides the solution. Test fixtures are cheaper than wafer probe stations and offer more flexibility to the RF port location. With a test fixture, the matching and bias circuitry can be located near the DUT. Yet test fixtures add parasitics, such as ground loops and RF reflections at the interfaces, to the characterization scheme. Since the calibration plane is not usually at the package's pads, accurate de-embedding becomes essential.

This chapter extends RF characterization principles developed for die characterization (see Chapter 6) to the package. It helps the reader understand the package's effect on the die, especially parasitic inductances (self, mutual, and ground) and capacitances (both self and mutual). The first part of this chapter explains how to design a package test fixture, followed by RF experiments for package characterization. Characterizing popular RFIC package styles is discussed at the end of the chapter.

8.1 Designing a Test Fixture for Package Characterization

A test fixture for package characterization has three basic components: the fixture's body, the RF launchers, and the DUT carrier. A sample assembly is shown in Figure 8.1. Package test fixtures are generally similar to other RF test fixtures. This section details each of the three components.

8.1.1 RF Launchers

The purpose of an RF launcher is to get RF signals on to and off of the fixture. At least three styles of launchers are available. The simplest is a coaxial connector soldered to the ends of a PCB (see Figure 5.1 in Chapter 5). The

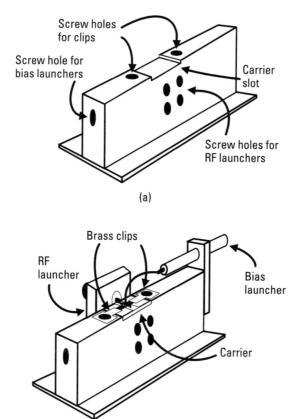

Figure 8.1 A test fixture designed for characterizing microwave and millimeter-wave packages: (a) the fixture's body, and (b) the fixture's body with the carrier and launchers inserted. For clarity, only one RF and bias launcher are shown.

coaxial center pin is soldered to a microstrip line on top of the PCB. Another choice is a replaceable RF launcher [shown in Figure 8.1(b)]. The launcher's coaxial center pin rests on top of the carrier's microstrip line, while the launcher's housing is screwed to the side of the fixture. At either end of the fixture is a coaxial bias line used to carry the dc, its center pin wire-bonded to the carrier. A third possibility is to design a carrier with probe pads laid out to accept coplanar probes. Regardless of the style of launcher, mounting it as close as possible to the DUT minimizes the RF effects of the transition.

For high-volume testing where new DUTs are constantly being measured, rugged and repeatable launcher contacts are essential. Spring-loaded RF contacts are particularly popular. However, the capacitance of the spring changes with pressure and, after repeated use, eventually wears out. Compliant "fuzz" buttons, which are long wires compressed into a tiny cylindrical shape, are also used in high-volume test fixtures. Another popular RF contact is the pogo pin, which resembles a miniature pogo stick, making pressure contact to the package's pads.

8.1.2   Coplanar Probes as RF Launchers

The transmission lines on the carrier can be designed to accept coplanar probes. CPW pads can feed a microstrip line leading to the package's RF ports (see Figure 8.2). The line length from the transition to the package should be de-embedded, setting the reference plane at the package's external pads (see Section 6.3 in Chapter 6).

Using coplanar probes to directly contact a package is only possible when the ground-signal pitch of the coplanar probe matches up with the package's pads. Usually, the package's ground pad is not near to or coplanar with a signal pad. One option is to use a CPW probe chip [1] (see Figure 8.3). Mount one CPW chip to the ground paddle inside the package and another one to a grounded area outside the package [2, 3]. Ground vias through the probe chip connect the coplanar grounds to the ground paddle inside the package. De-embedding the via's ground inductance gives the best accuracy.

8.1.3   Test Fixture Body

Typically machined from aluminum, the test fixture body is the block that holds together the entire assembly. Referring to Figure 8.1, RF and bias launchers are screwed to its sides while brass clips secure the carrier in the slot on top. To electrically define the reference plane, the carrier can be replaced with calibration standards.

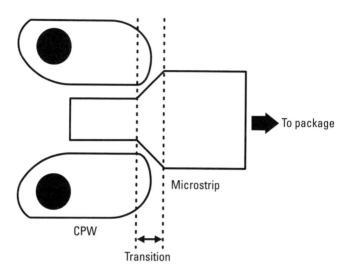

Figure 8.2 A coplanar-to-microstrip transition. The design of the transition region becomes more important as the frequency increases.

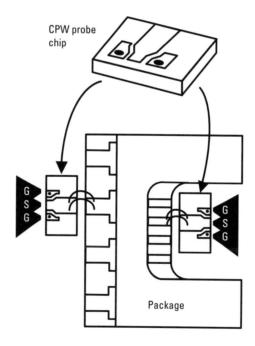

Figure 8.3 Using coplanar probe chips to characterize a package. Shown are G-S-G probes. The two probe chips are used to characterize the package's feedthrough path.

A special test fixture design is used for characterizing vertical packages. Vertical packages have one RF contact on top and the other on bottom [4]. The RF signal travels vertically through the packaged device. Diodes are a good example. A diode test fixture has a barrel shape to adapt to the diode's radial field pattern (see Figure 8.4). By applying the L-matrix and C-matrix techniques (explained in Section 8.6.6), a vertical fixture can be calibrated by inserting a shorted package (L-matrix) and an open package or no package inserted (C-matrix) [5].

8.2 The Carrier

For easy insertion and removal in the fixture, the package and interconnecting lines are mounted to a carrier (see Figure 8.5). An assembly that includes the launchers (less the fixture body) is shown in Figure 8.6. Designed for

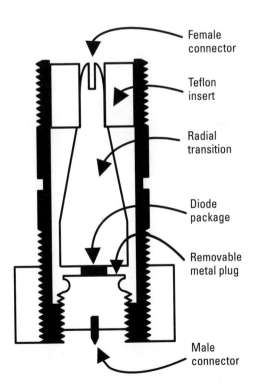

Figure 8.4 A vertical test fixture, designed for leadless or "pill"-type diode packages.

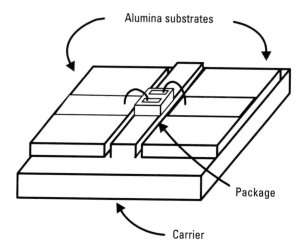

Figure 8.5 The package mounted to a copper carrier. Two alumina substrates hold input and output microstrip lines.

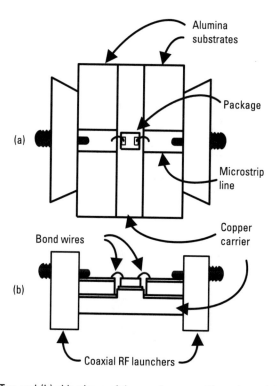

Figure 8.6 (a) Top and (b) side views of the carrier assembly and coaxial RF launchers. The package can be grounded by bonding to the center of the carrier.

millimeter-wave frequencies, it uses alumina substrates affixed to a gold-plated copper carrier. At RF frequencies, the carrier and substrates can be a single PCB rather than an assembly. In either case, the carrier can contain bias and impedance-matching circuitry along with the RF microstrip lines.

Whether the substrates are organic (as with PCB) or ceramic (such as alumina), their TCE should be as close as possible to that of the carrier and package. Otherwise, bonding the package to the substrate will induce mechanical stress. Such stress is transferred to the package and die more so than it is to the substrate. When stress reaches the die, it can even go so far as to affect the carrier mobility within the semiconductor [6].

8.2.1 Designing the Carrier

The following are a few points to consider when designing the carrier and substrates. Via "fences" on either side of a transmission line isolate it from adjacent lines (see Figure 8.7). The via hole spacing S should be no closer than three times the substrate height h; otherwise, radiation and coupling

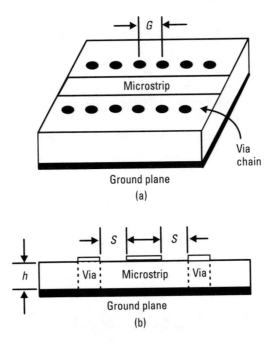

Figure 8.7 (a) Via fences isolating a microstrip line and (b) the side view. G is the separation between ground vias.

occur [7]. To suppress this, as well as higher-order modes along the line, the via post separation G should not be more than $3h$. Connecting the vias on the top surface with a solid conductor line further improves isolation but can induce CPW modes, corrupting the desired microstrip mode.

A two-layer laminated PCB works well for operation up to a few GHz. The top and bottom metal layers can contain ground planes with a power plane in the middle for the conductor layer [8]. Using ground planes on the top and bottom provides good shielding, which lowers the RF line loss. Be sure to connect the ground planes, otherwise a potential difference between them will form capacitance. Unconnected, the parallel ground planes behave as a magnetic wall at the board edges. By shorting the top and bottom ground planes at the edges, the parallel ground planes instead act as an electric wall, creating resonance in the measurement. The solution is to connect the top and bottom grounds with vias intermittently throughout the grounded region to suppress such resonances.

8.2.2 Carrier Board Material

In general, PCB is used at RF frequencies and alumina substrate at microwave and millimeter-wave frequencies. There are reasons for avoiding PCB at higher frequencies. For one, PCB is a soft material. The characteristic impedance Z_0 of a line is a function of board thickness, so uneven thickness causes the line's Z_0 to vary along its length. The conductor plating on a PCB is etched in a wet chemical process, so that line width control is not as good. The PCB's dielectric surface is rough, having an impact at millimeter-wave frequencies. Also, the dielectric filling within the PCB is not as consistent or as uniform as alumina.

At higher frequencies, alumina has better RF qualities than PCB. It is a hard, ceramic material whose surface can be highly polished for smoothness. For example, 98.6% alumina has a high dielectric purity and good dielectric uniformity. The conductors are deposited in a photolithographic process for better line-width control. The disadvantages are its poor thermal conductivity, brittleness, and high cost. Alumina's higher dielectric constant also causes higher microstrip loss than with PCB.

Mutual coupling commonly occurs between the package and carrier, altering how the package is modeled. A few qualities of the carrier and substrate will determine the amount of coupling, such as the uniformity of the substrate's dielectric, its hardness, the conductor plating, and, most importantly, how they are attached.

8.3   Attaching the Package to the Carrier

8.3.1   Bond Wires

Bond wires are an easy way to connect RF lines from the carrier substrate to the package. This section discusses the RF implications of using bond wires.

8.3.1.1   Bond Wire Inductance

When the bond wire is short compared with a wavelength, its equivalent circuit is simply a lumped inductance with a small series resistance. By definition, inductance L is the proportion of magnetic flux ψ to the current I flowing through a wire. In a bond wire, inductance exists both internally and externally (see Figure 8.8). At dc, I flows uniformly throughout the bond wire. As the frequency f increases, I flows closer to the surface. This property is known as the skin depth δ

$$\delta = \sqrt{\frac{1}{\pi f \mu \sigma}} \tag{8.1}$$

where f is the frequency (Hz), μ is the permeability (H/m), and σ is the conductivity (mhos/m).

As a rule of thumb, the skin depth for bulk gold as a function of frequency is

$$\delta = \frac{0.075}{\sqrt{f}} \tag{8.2}$$

As f increases, the current moves toward the surface. As a result, the bond wire's internal inductance L_{int} decreases as the square root of f or -10 dB/decade. Externally, magnetic flux lines ψ run just beyond the wire's diameter, leading to external inductance L_{ext}. A bond wire exhibits self-inductance along with mutual inductance to neighboring conductors. A dynamic RF current I will create ψ nearer the surface, too. As I crowds to the wire's surface, less of it flows due to skin effect, making for less ψ.

As the RF current cycle increases and decreases sinusoidally, the magnetic field surrounding a bond wire expands and contracts, inducing a voltage. Each bond wire's self-inductance is the effect of ψ on the bond wire itself. Mutual inductance is how ψ induces a voltage in a nearby conductor. Mutual inductance rapidly decreases with an increasing separation between conductors.

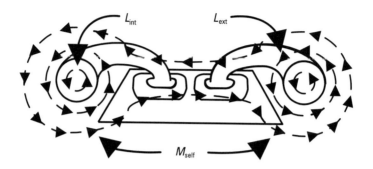

- Internal bond wire inductance L_{int}
- External bond wire inductance L_{ext}
- Mutual coupling M_{self}
- $L_{self} = \Sigma(L_{int} + L_{ext})$ each bond wire $+ \Sigma(M_{self})$ all bond wire combinations

Figure 8.8 Cross-section of two bond wires mounted to a bond pad. Also shown are the magnetic fields (dotted lines) and the self-inductances (internal, external, and mutual). L_{self} is the total self-inductance of the bond wire pair.

If two parallel bond wires connect to the same pad, then obviously current flows in both. Proximity of the bond wires to one another causes M_{self} to increase due to the magnetic fields, adding instead of canceling as they do with the fields induced by the ground current. Since I for the two bond wires flows in the same direction, external ψ lines constructively couple between the two. The net self-inductance L_{self} of the connection is the sum of the individual bond wire inductances $(L_{int} + L_{ext})$ plus the mutual coupling M_{self} between them. When bond wires are orthogonal, no coupling M_{self} occurs.

This analysis does not consider what is occurring with the ground plane. The ground plane contains an image current flowing in an opposite direction to the signal current (see Section 8.7). The image current emits magnetic fields counter to those of the signal conductor, ideally canceling with them. When the bond wires are close to the ground plane, the L_{ext} and M_{self} of the image current cancels with the L_{ext} and M_{self} of the bond wires [9]. In general, having the ground plane nearby results in smaller values of L_{ext} and M_{self}.

In practical circuit design, a bond wire's series inductance can be tuned with the shunt capacitance of a bond pad. Such capacitance can be created by widening the bond pad where the bond wire connects to it [10]. At millimeter-wave frequencies, the wire bond becomes a significant fraction of a wavelength, behaving as an element in a filter network. Hence, filter theory can be used to properly tune the bond wire-to-bond pad connection [11].

8.3.1.2 Bond Wire Capacitance

Capacitance exists from the bond wire to the ground. At frequencies above 6 GHz, both substrate and mutual capacitances arise (see Figure 8.9) [12]. In general, the farther the bond wire is from the substrate, the less capacitance exists. The amount of bond wire-to-bond wire capacitance rests on the potential difference between them. The curvature of the bond wires will affect the self and mutual inductances, as well as the capacitance to the substrate [13].

8.3.1.3 Bond Wire Resistance

The bond wire's geometry (i.e., its length and curvature) can affect its high-frequency resistance, so employing electromagnetic simulator software is the best predictor of resistance values. As a reference, the high-frequency resistance of a straight wire is

$$R_{ac} \cong \frac{l}{2\pi r \delta \sigma} \tag{8.3}$$

where δ is the skin depth, σ is the conductivity, r is the radius of the wire, and l is its length.

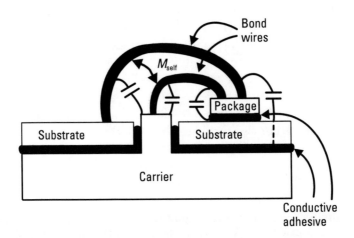

Figure 8.9 Parasitic capacitances associated with the package and substrate. Also shown is the mutual inductance M_{self} between the bond wires.

8.3.2 Conductive Adhesives

In Figure 8.9, the package and substrates are attached with an electrically conductive adhesive. Electrically conductive adhesives can be used in place of solder, serving as both a mechanical bond and an electrical connection. Some advantages of using a conductive adhesive include the following:

- Will not leach the gold from the conductor line like solder;
- Can be used on nonsolderable surfaces;
- Can be applied at a lower temperature;
- Has a lead-free content;
- Requires no flux or flux cleaning after assembly.

The two classes of conductive epoxy are *anisotropic electrically conductive adhesives* (ACAs) and *isotropic electrically conductive adhesives* (ICAs). ACAs have conductive particles mixed nonuniformly throughout a resin filler. In ICAs, the conductive particles are evenly distributed. In either case, the size of the conductive particles can vary.

A simple way to characterize the RF performance of a conductive adhesive is by bridging a gap in a microstrip line [14, 15]. A short length of 50-Ω line is bonded across the gap using the adhesive (see Figure 8.10). Because the RF effect of the adhesive can be small compared with the line, a low-loss insulating dielectric such as GaAs should be used as the bridge substrate. Otherwise, the RF performance of the substrate and the bridge will dominate the measurement. A calibration method such as TRL can be used to define the reference plane at the dotted lines in the figure.

There are drawbacks to this technique. To gauge the epoxy's effect, a soldered thru can be used as a reference. Care should be taken during assembly of the bridge. Otherwise, the procedure will measure not only the quality of the adhesive but also the repeatability of the assembly process. Note that the reference planes slice vertically through the substrate. The contribution of the substrate area underneath the bridge will be included as part of the adhesive's RF performance. The coupling of the fields from the bridge to the substrate will change when underfill is applied.

A similar characterization method relies on the change in quality factor Q of a bridged $\lambda/2$ resonator (see Figure 8.11). An epoxied 50-Ω line again bridges a gap in the microstrip [16]. Designing the $\lambda/2$ resonator for weak coupling, it should have a loaded Q_L of

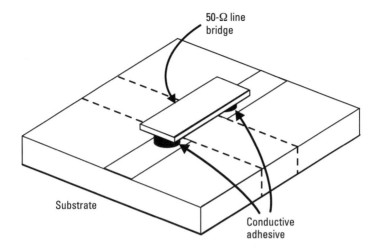

Figure 8.10 Characterizing the RF performance of a conductive adhesive. A 50-Ω line is mounted upside down on a substrate. Before measuring, the reference planes should be calibrated to the dotted lines.

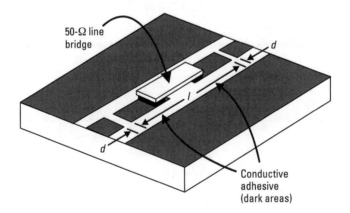

Figure 8.11 End-coupled λ/2 resonator jumpered in the center with a 50-Ω line. The *d* determines the amount of resonator coupling. The resonator length *l* should be λ/2 of the center frequency.

$$Q_L = \frac{f_r}{\Delta f} = \frac{\omega L_s}{R_s} \tag{8.4}$$

where f_r is the resonant frequency, Δf is its 3-dB bandwidth, R_s is the series resistance, and L_s is the series inductance of the resonator. The amount of deviation of Q_L from an unbridged $\lambda/2$ resonator quantifies the high-frequency contribution of the epoxy's L_s and R_s.

The RF equivalent circuit of the adhesive depends on how it was applied. The top of Figure 8.12 shows the case of high surface tension as the adhesive spreads out on the carrier. This makes the capacitance larger close to the carrier's surface. The middle of Figure 8.12 illustrates a bulge due to less surface tension. This can be due to either a lack of flux or an excessive amount of adhesive. Electrically, the bulge results in more capacitance to the carrier and package. The bottom of the figure shows a lack of coverage where not enough adhesive is applied or perhaps the carrier has solder flux. These three cases show how the amount of adhesive dispensed from the bonding machine results in a variance of the capacitance to the carrier.

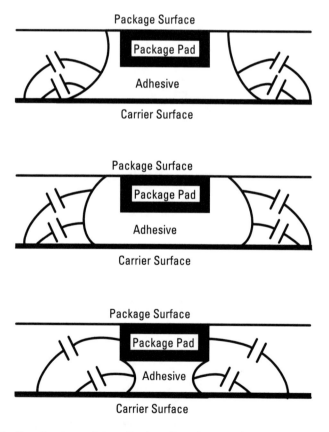

Figure 8.12   How the shape of the adhesive affects its capacitance.

8.4 Calibration

Regardless of the style of RF launcher used on the test fixture (i.e., coaxial connector and coplanar probe), it must be de-embedded from the measurement. The best way to set the reference plane at the DUT's ports is by mounting calibration standards in place of the package (see Figure 8.13). The difficulty with this approach is that the standards must be well characterized a priori, particularly troublesome with short and load standards that use vias to ground with inductance. Calibration using TRL is preferred over SOLT because the parasitics of SOLT standards mounted in a fixture are difficult to determine [17].

Another way to de-embed is to model the launcher and microstrip interconnect lines by their equivalent circuits. The problem is that at higher frequencies, the equivalent-circuit model of the launch becomes complicated. The simplest, but least accurate, de-embedding method is to insert a

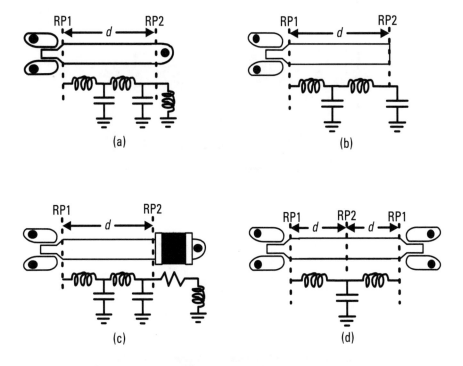

Figure 8.13 SOLT standards using coplanar probe launches: (a) short, (b) open, (c) load, and (d) thru. Calibration shifts the reference plane from the launcher plane (RP1) to the package ports (RP2).

thru line in place of the DUT and use matrix math to de-embed the fixture (see Section 8.5.1).

8.4.1 Partitioning by Reference Planes

During test, the die, its package, and the fixture's carrier substrate all act as a single RF assembly (see Figure 8.5). Discerning the individual RF behavior of each can be found by defining reference planes. Figure 8.14 shows a block diagram of the combined assembly. An error model represents each section,

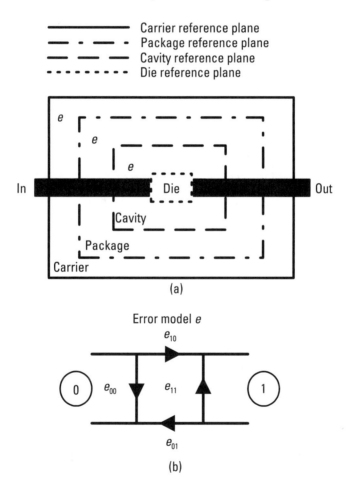

Figure 8.14 (a) Block diagram dividing the carrier and DUT assembly into reference planes. (b) Each section can be represented by an error model e.

starting at the carrier and leading to the die. Error models are useful in finding the amount of error each section contributes to the overall measurement.

In reality, sectioning by reference planes is challenging (see Figure 8.15). For instance, a cavity can exist in the package that contains the die and bond wires. Coupling occurs from the bond wires to nearby leads in the package, as well as to the die. This coupling makes the cavity difficult to characterize. Arbitrarily allotting coupling effects across reference planes is valid. However, such decisions should be clearly spelled out.

To account for the RF effects that the ground paddle adds to the die, try modeling the paddle as a separate, distinct section in the RF assembly (see Figure 8.16). Schematically connecting to the die with transformers forces the current running through the die to behave in accordance with its actual electromagnetic behavior [18, 19]. Specifically, the current injected into the signal path's bond wires returns not by way of the ground bond wires but through the image current in the ground paddle directly underneath. Connecting to the die through a transformer also makes circuit simulations of the die easier to insert into a package [20].

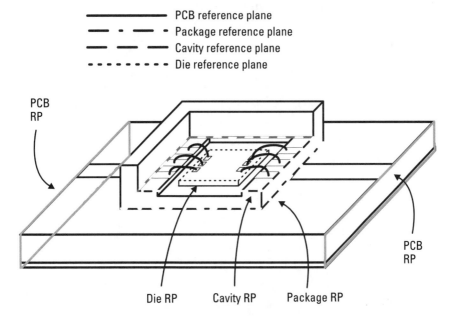

Figure 8.15 Physically sectioning by reference planes. A packaged die is mounted to a PCB carrier substrate.

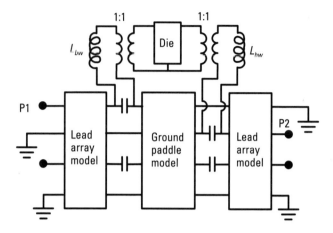

Figure 8.16 Partitioning the package's ground paddle to better account for its electromagnetic effects. Using transformers permits separate modeling of the die and the package. P1 and P2 are the RF ports and L_{bw} is the bond wire inductance.

8.5 De-embedding the Fixture from the Measurement

A number of schemes are available to de-embed the fixture's effects from the measurement, namely de-embedding the microstrip lines on the PCB in Figure 8.15. These lines can be de-embedded in one of the following ways:

- *Linear phase compensation.* Adding a phase delay can compensate for the phase due to a length of transmission line. This technique assumes that the phase is linear and that the line is lossless (see Section 2.5.1 in Chapter 2).

- *Normalization.* A short and a thru can be used to normalize an in-fixture measurement. To find the return loss of the DUT ($S_{11,DUT}$), normalize a return loss measurement (S_{11M}) by inserting a short in place of the DUT (see left side of Figure 8.17). The short (whose S_{11} = –1) finds the reflection tracking error E_{RF}

$$S_{11}^{Short} = E_{DF} + \frac{(S_{11,DUT})(E_{RF})}{1-(E_{SF})(S_{11,DUT})} = E_{DF} + \frac{(-1)(E_{RF})}{1-(E_{SF})(-1)} \cong -E_{RF} \quad (8.5)$$

- Note that a reflection normalization does not correct for directivity (E_{DF}) or source mismatch errors (E_{SF}) but instead assumes they are zero (see Figure 2.4 in Chapter 2). Similarly, replacing the DUT with a thru gives the transmission error E_{TF} (see right side of Figure 8.17). Transmission normalization does not correct for factors such as the load mismatch error E_{LF}.

- *In-fixture calibration.* Mount calibration standards one by one in place of the DUT and calibrate. When quality standards are available, this is the best choice.

- *Thru line.* Replace the DUT with a thru line that connects the input and output lines on the carrier substrate (see Section 8.5.1).

- *Time domain gating.* Transform the S-parameters from the frequency domain to the time domain, then bound the discontinuity in time. Transforming back to the frequency domain gives the frequency response of the discontinuity (see Section 8.5.2).

8.5.1 De-embedding by Using a Thru Line

The easiest de-embedding method to perform is to replace the DUT with a 50-Ω thru line. The thru line should be the combined lengths of the RF

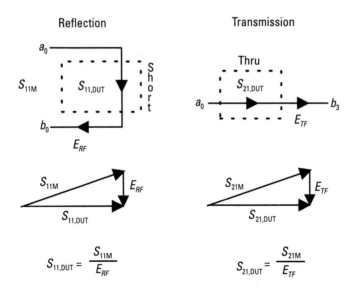

Figure 8.17 Normalizing the reflection (port 1 return loss) and transmission (insertion loss) measurements, inserting a short and a thru in place of the DUT, respectively.

input and output lines (see Figure 8.18). For scalar de-embedding, simply measure the insertion loss of the thru and subtract its magnitude from the DUT measurement.

When the fixture is symmetrical (i.e., port 1 and port 2 are interchangeable), then the input and output lines can be de-embedded from the measurement using matrix algebra [21, 22]

$$[A_{MEAS}] = [A_{IN}] \bullet [A_{DUT}] \bullet [A_{OUT}] \qquad (8.6)$$

where the measured ABCD matrix $[A_{MEAS}]$ is composed of the input line $[A_{IN}]$, the output line $[A_{OUT}]$, and the DUT $[A_{DUT}]$ matrices. With symmetry, $[A_{IN}]=[A_{OUT}]$, so that the DUT matrix can be mathematically de-embedded from the measurement. When there is not symmetry, inserting several different line lengths and an open in place of the DUT can serve to de-embed the input and output lines in a similar fashion [23].

8.5.2 Time-Domain Analysis

Injecting a pulse into a port and studying its time-domain response can reveal RF behavior within the DUT. Alternatively known as *time-domain*

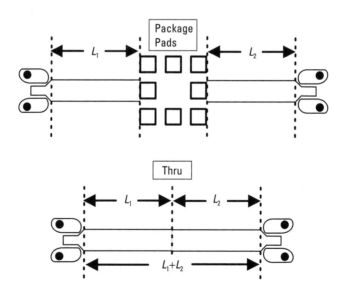

Figure 8.18 A thru used to de-embed the microstrip lines leading to the DUT. The thru is the combined length of the input (L_1) and output (L_2) lines.

reflectometry (TDR) or *time-domain transmission* (TDT), time-domain analysis uncovers RF behavior at points beyond the calibrated reference plane [24]. Time-domain *reflectometry* is the study of the pulse reflected from the DUT, while time-domain *transmission* looks at the pulse traveling through the DUT.

The time delay of the returning signal correlates to a physical distance traveled. The shape of the returning pulse tells much about discontinuities within the DUT beyond the coaxial calibration plane. This ability can be handy in understanding the behavior of test fixture mismatches and transitions. The returning pulse's shape can show whether the fixture mismatch is capacitive or inductive (see Figure 8.19). If the returning pulse has the same shape but simply attenuated, then the load presented by the DUT is resistive.

Using an oscilloscope to examine the shape of the reflected pulse is the traditional method of time-domain analysis. When equipped with a time-domain option, modern VNAs can take S-parameters and Fourier transform them into a time-domain waveform. Either TDR or TDT can be used to find the distance to the discontinuity. Distance resolution is inversely proportional to the frequency span of the measurement. The wider the measurement bandwidth, the smaller the distance or time discernable. Overwidening the bandwidth can inadvertently encompass higher-order modes that will distort the results.

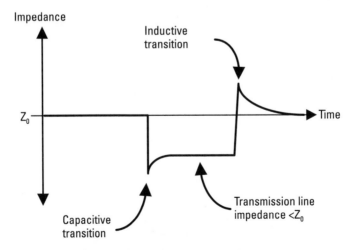

Figure 8.19 The time-domain pulse shape reveals the nature of an impedance mismatch beyond the calibration plane.

When impedance discontinuities are close together, isolating them spatially can be difficult. Energy reflected from one discontinuity lessens the amount of energy continuing on to the next discontinuity. The discontinuity closest to the calibrated reference plane will impact all the following discontinuities. When one discontinuity clouds the behavior of another discontinuity, this is called *masking*. *Response resolution* is how close together two discontinuities can be located and still be able to pick out their individual electrical behavior. How close two discontinuities can be in time or space while still electrically discerning them is called *range resolution.*

Placing bounds in the time domain around a discontinuity is referred to as *gating* (see Figure 8.20). With gating, discontinuities in the line can be isolated and de-embedded from the measurement. Building a rectangular box in the time domain equates to a $\sin(x)/x$ function in the frequency domain. Note that the $\sin(x)/x$ function has sidelobes in the frequency domain that show up as ringing in the time-domain sweep.

To be able to discern impedance discontinuities, calibrating TDR and TDT measurements is crucial [25]. Begin either TDR or TDT by performing a standard calibration (i.e., SOLT and TRL) in the frequency domain to define the reference plane. To calibrate for TDR, place a reflect (such as a short) a premeasured distance from the reference plane. The input voltage,

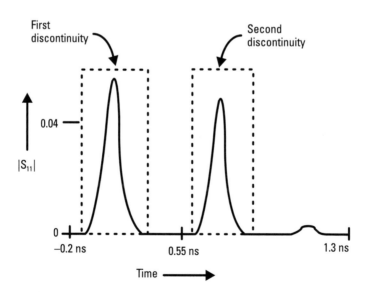

Figure 8.20 Use TDR with a return loss measurement to gate discontinuities. Usually discontinuities are gated one at a time.

$V_{REF}(t)$, and the voltage returning from the DUT, $V_{DUT}(t)$, have a delay that reveals the round-trip propagation time for the pulse. Fourier transforming F the two gives the return loss $S_{11}(\omega)$

$$S_{11}(\omega) = -\frac{F[V_{DUT}(t)]}{F[V_{REF}(t)]} \tag{8.7}$$

where $V_{DUT}(t)$ is the measured voltage pulse returning from the DUT and $V_{REF}(t)$ is the input voltage.

A simple method to calibrate for TDT is to use a thru to find the propagation time from port to port. The thru should be long enough so that reflections from the connectors do not interact with reflections from the DUT interface (see Figure 8.21). The thru's insertion loss is the ratio of the voltage reflected from the DUT, $V_{DUT}(t)$, to that transmitted through the thru $V_{THRU}(t)$. Fourier transforming each from the time domain to the frequency domain gives

$$S_{21}(\omega) = -\frac{F[V_{DUT}(t)]}{F[V_{THRU}(t)]} \tag{8.8}$$

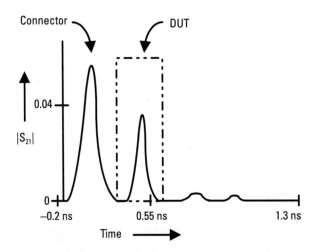

Figure 8.21 Using TDT to gate the DUT's response and discern it from the effects of the connector.

8.5.3   How to Apply Time-Domain Techniques

To measure a DUT, the RF input and output (I/O) lines leading to the DUT must be de-embedded. Gating the RF connectors and I/O lines separately can be difficult because their close proximity causes them to mask one another. To de-embed the connectors and I/O lines, use two standards, a thru and an open, respectively [26, 27]. First insert a thru in place of the DUT, gating the I/O connectors in the time domain. To view their S-parameters, Fourier transform back to the frequency domain. The thru should be long enough to isolate the I/O connectors from one another, but short enough that the thru's length is not long and lossy. To de-embed the RF I/O lines, insert an open. The I/O lines leading to the open should be the same lengths and design as those leading to the DUT. Gate each I/O line individually, starting the gate at the connector plane found earlier and stopping at the open. Then Fourier transform the line's measurement to S-parameters. The S-parameters can be used in an ABCD chain matrix fashion to de-embed the connectors and I/O lines individually and move the reference plane to the DUT's ports [see (8.6)].

To gauge physical distance on a time-domain display, probe the I/O line using a tuning stick. The tuning stick is simply a toothpick with copper foil on its tip. By probing the line with a toothpick, the copper foil introduces discontinuities along the line. Looking at the display, identify the stick's location on the time-domain trace. This can help place a gate around a discontinuity.

8.6   Procedure for Characterizing a Package

Having developed an understanding of the test fixture's construction (see Sections 8.1 to 8.3), its calibration (Section 8.4), and fixture de-embedding (Section 8.5), we are now ready to characterize a package.

Consider a package with eight leads (see Figure 8.22). Six of them are routed through the package walls to bond pads inside, while the other two go to a ground paddle in the package's center. For the sake of clarity, the internal routing through the package walls is not shown. The task is to generate a high-frequency equivalent-circuit model of the package.

Begin by sketching an equivalent circuit, looking at the package's physical construction. Based on how the leads run through the package and its dielectric layers, the circuit may consist of resistances, capacitances (both self and mutual), and inductances (both self and mutual). Estimate rough values for the elements of the equivalent circuit.

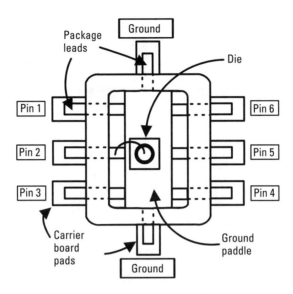

Figure 8.22 A common 8-pin package. The die is wire-bonded internally to bond pad 2.

Next connect the RF test ports to the package. The package shown has six RF ports, but a VNA has only two ports. Even if more test ports are available, modeling all the pins at once would be too complicated [28]. Characterize all combinations of pins two at a time, using symmetry wherever possible.

Before beginning any measurements, a necessary and sufficient set of experiments should be outlined to evaluate each circuit path in the package. For instance, consider adjacent pins 1 and 2 (see Figure 8.23). The external leads are labeled P1 and P2, while the internal wire bond pads are P1′ and P2′. The first experiment is to make two-port S-parameter measurements with the internal pads P1′ and P2′ unconnected or open [see Figure 8.23(a)]. At low frequency, the capacitance terms dominate so that the inductive circuit elements can be X'ed out. The resistors represent RF leakage through the resin dielectric of the package wall. Because no current flows, the resistors can be X'ed out, too. For the next experiment, short P1′ and P2′ internally with a bond wire to the ground paddle [see Figure 8.23(b)]. Because the inductances are a larger part of the impedance at high frequency, the capacitors can be X'ed out. To isolate and identify coupling between pins 1 and 2, two experiments are a thru [see Figure 8.23(c)] and, for a one-port S-parameter measurement, a grounded thru [see Figure 8.23(d)]. The ground paddle is represented as a simple inductance L_g and capacitance C_g.

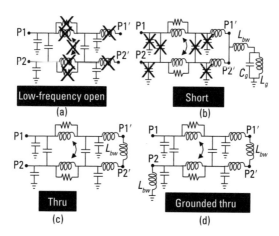

Figure 8.23 Circuit configurations used to model adjacent leads in a package: (a) low-frequency open, (b) short, (c) thru, and (d) grounded thru. P1 and P2 are the external package leads and P1 and P2 the internal bond pads. Also shown is the bond wire inductance L_{bw}.

Higher modeling precision may require a more complicated circuit [18]. In all experiments, a circuit simulator helps to fit the measured data to the model's equivalent-circuit values.

Creating a lumped-element equivalent circuit of the package such as the one shown in Figure 8.23 is valid as long as the one-way RF path length through the package is less than a tenth of a wavelength ($\lambda/10$). One way to determine the path length is by measuring the phase of the port's return loss. If the phase of the return loss (S_{11} or S_{22}) is greater than 0.4 radians (where $0.4\pi = 2$ (for round trip) • 2π (for a wavelength) / 10), then the path length is greater than $\lambda/10$ and a more complicated equivalent-circuit model is needed [29].

8.6.1 Characterizing with a Thru

A simple thru connection inside the package can be made using a bond wire from P1′ to P2′ [see Figure 8.23(c)] [30]. Measuring a thru's insertion loss will reveal not only the self-inductance L_{self} and self-capacitance C_{self} of each path but also the mutual coupling M between them. When measuring, note that the bond wire itself can couple to neighboring pins and the ground paddle.

Instead of a bond wire, another possibility for a thru is mounting a die on the ground paddle that contains a 50-Ω thru line. For best results, the dimensions and properties of the thru die should be as close as possible to the actual DUT die.

8.6.2   Characterizing with a Short

The simplest way to short a path through the package is to connect the bond pad inside the package to the ground paddle. A better solution is to use a shorted die or a solid conductor in place of the die. In this way, the arc and length of the bond wire will resemble that of the actual die. The accuracy of this approach depends on how well the shorted die resembles the actual die. Similarly for the open, a bond wire connecting from the bond pad to an open-circuited die is preferable.

8.6.3   Characterizing with an Open

Per Figure 8.23(a), measuring the return loss of an open will reveal the capacitances (both self and mutual) of a single path. Since an open does not cause current flow, the series resistance, series inductance, and mutual inductance of the path are negligible.

There are disadvantages to the short and open methods discussed in Sections 8.6.2 to 8.6.3. With an open or a short nearly all the incident energy is reflected back. In either case, the VNA computes small differences between the forward and reflected waves. As the ratio of the forward-to-reflected waves gets smaller, the number of significant digits in the computation eventually reaches the resolution of the VNA (the noise floor). Instead of using a short or open, terminating the bond pad inside the package in the path's characteristic impedance Z_0 is the ideal solution. This would present a manageable difference between the forward and reflected waves. However, finding the path's Z_0 and building a load to match Z_0 is seldom practical. Instead, consider using a 50-Ω termination.

8.6.4   Characterizing with a Load

Terminating the path with a 50-Ω load inside the package and measuring the return loss will yield the series resistance, the series inductance, and the shunt capacitance to ground of the path being studied. To terminate, bond wire a 50-Ω chip resistor to an internal pad inside the package [31]. One end of the resistor can be connected either to the ground paddle or to an internal pad that is grounded outside of the package. In either case, the inductance of the ground should be included during modeling.

When two package paths are simultaneously terminated with separate loads, measuring the insertion loss between the two unconnected paths will reveal their mutual inductance and mutual capacitance.

8.6.5   Characterizing with a PIN Diode

Another technique to characterize each package path is to mount an element inside whose impedance can be changed electronically [32]. Using a PIN diode as a characterization element, adjust the bias to change the impedance presented to the internal port (see Figure 8.24). A PIN diode appears as a large capacitance in its "off" (unbiased) state and a small resistance in its "on" (forward-biased) state.

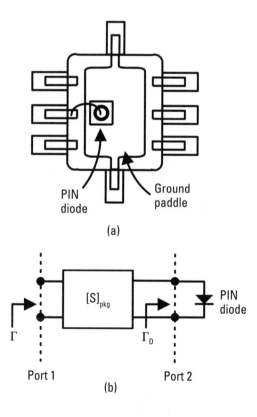

(a)

(b)

Figure 8.24 (a) A PIN diode mounted inside a leaded package; and (b) the equivalent circuit of the RF path. $[S]_{pkg}$ are the S-parameters of the path through the package.

In Figure 8.24(b), port 1 is the external package pad and port 2 is the wire-bonded contact on top of the diode. To find the reflection coefficient of the package path between port 1 and port 2, apply the equation

$$\Gamma = S_{11} + \frac{\Gamma_D S_{21} S_{12}}{1 - S_{22} \Gamma_D} \tag{8.9}$$

where Γ_D is the reflection coefficient of the PIN diode at port 2, found by characterizing the PIN diode before mounting it into the package.

The difficulty with this technique is that port 2 inside the package cannot be accessed easily. A workaround is to measure the 1-port S-parameters at the package's external port (port 1), biasing PIN diode to three impedance states (a,b,c). These three impedance states yield three different reflection coefficients Γ_a, Γ_b, Γ_c. Assuming $S_{21} = S_{12}$, the two-port S-parameters of the package in Figure 8.24 can then be found by

$$S_{11,\text{pkg}} = \frac{-\left(\Gamma_a \Gamma_b \Gamma_{D,c} \Delta_1 + \Gamma_b \Gamma_c \Gamma_{D,a} \Delta_2 + \Gamma_a \Gamma_c \Gamma_{D,b} \Delta_3\right)}{\Delta} \tag{8.10}$$

$$S_{22,\text{pkg}} = \frac{\Gamma_c \Delta_1 + \Gamma_a \Delta_2 + \Gamma_b \Delta_3}{\Delta} \tag{8.11}$$

$$S_{21,\text{pkg}} S_{12,\text{pkg}} = \frac{(\Gamma_a - \Gamma_b)(\Gamma_b - \Gamma_c)(\Gamma_c - \Gamma_a)\Delta_1 \Delta_2 \Delta_3}{\Delta^2} \tag{8.12}$$

where

$$\Delta = \Gamma_c \Gamma_{D,c} \Delta_1 + \Gamma_a \Gamma_{D,a} \Delta_2 + \Gamma_b \Gamma_{D,b} \Delta_3 \tag{8.13}$$

$$\Delta_1 = \Gamma_{D,a} - \Gamma_{D,b} \tag{8.14}$$

$$\Delta_2 = \Gamma_{D,b} - \Gamma_{D,c} \tag{8.15}$$

$$\Delta_3 = \Gamma_{D,c} - \Gamma_{D,a} \tag{8.16}$$

As mentioned, the reflection coefficients $\Gamma_{D,a}$, $\Gamma_{D,b}$, and $\Gamma_{D,c}$ of the PIN diode at impedance states a, b, c are best measured beforehand on-wafer. The three impedance states chosen for the PIN diode should be spread evenly across the Smith chart, ideally emulating an open, a short, and a load.

Mounting PIN diodes on several leads at the same time would permit full characterization of each package path, including their mutual coupling.

8.6.6   What to Do with the Unused Package Pins

The question arises of how to deal with the unused pins, that is, the package leads not connected to the RF test ports. Deciding whether they are grounded or left open (unconnected) will affect the RF response at the test ports. Generally, open pins increase their mutual capacitance to the pins-under-test [33]. Conversely, grounding the neighboring pins to the grounded die paddle inside the package will increase the amount of M. Consider the case of unused pins that are grounded both inside and outside the package. At frequencies up to several hundred megahertz, the mutual capacitance from the test lead to a shorted pin is much smaller than its mutual inductance. Hence, the equivalent-circuit model to a neighboring grounded pin is principally inductive (both self and mutual). In the tenths of a Ω range, the resistance of a grounded test lead is the same order of magnitude as the contact resistance of an RF launcher, which varies with each connection.

Using circuit theory, the L- and C-matrix methods illustrate the benefit of leaving pins open or shorted [34]. Look at pin 1 (P1) and pin 2 (P2) with the two pads grounded inside [see Figure 8.23 (b)]. All the other unused pins are grounded both internally and externally. When grounding all pins, the capacitances are small, making the self and mutual inductances and the resistances of pins 1 and 2 easier to find. This technique is known as the L-matrix method. The inductance of the ground paddle by itself can be found by removing the grounded bond wires inside from the unused pins one at a time. The increase in the ground inductance as each bond wire is pulled is proportional to the number of unused pins that are still grounded. The incremental increase in inductance accounts for the just-opened path. The unchanged, fixed amount is that of the ground paddle.

Now consider when all the unused pins are open-circuited both inside and outside the package. By open-circuiting the unused pins both internally and externally, the inductances can be ignored and the self and mutual capacitances isolated in a manner similar to the L-matrix method. This technique of open-circuiting and studying the change in capacitance is known as the C-matrix method [35].

8.7   Modeling a Package Mounted to a Carrier

The design of the carrier (i.e., its layout, board thickness, and dielectric) undoubtedly affects the RF behavior of the package. Before discussing how to de-embed the carrier's effect, a few fundamental questions should be asked. Since any package must be mounted to a carrier board to use it, should not the carrier and package be modeled together? Or should a package model be developed separately, independent of the carrier style? While modeling the package separately may seem more desirable, its RF behavior will always depend on the nature of the carrier. The layout of the carrier substrate will differ from one application to another, so modeling the package as though it were floating in free space is not only unrealistic, it can give misleading results when attached to one carrier design or another. Excluding the effect of the carrier and the attach material (either a conductive adhesive or solder) makes the package model more universal but also makes deriving such a model and applying it to a design more difficult.

8.7.1 Ground Inductance

Establishing a dependable RF ground path to the die inside the package is crucial. Regular insertion and removal of the DUT into a test fixture can make repeatable grounding performance difficult. For proper current conduction, the test circuit (die, package, carrier, test fixture, and test system) should form a common-mode current loop (see Figure 8.25). With a single ground, the signal current I_{sig} and ground current I_{gnd} are equal and flow in opposite directions. The magnetic fields (dotted lines) will cancel, and the electric fields (solid lines) will terminate properly. When the carrier is mounted in the fixture's slot (see Figure 8.1), a solid ground connection must be made along the bottom of the carrier to the fixture's slot. The less contact across the two surfaces, the more discrete ground points exist, raising the ground plane inductance.

Rarely is a single ground path made, rather multiple ground points occur in the circuit. This gives rise to differential mode currents (see Figure 8.26). Multiple grounds can cause ground current I_{gnd} to flow in the same direction as the signal current I_{sig} at points, leading to mutual coupling M and radiation.

Multiple ground paths set up ground currents traveling in different directions. Their effect shows up in the measurement as high-frequency resonance and generally higher crosstalk. In particular, ground discontinuity effects will disturb sensitive RF devices such as low-noise amplifiers and filters. In Figures 8.25 and 8.26, the ground plane image current runs along the side of the conductor plating facing the substrate. The plating conforms to the contours on the surface of the substrate. This is why, at millimeter-wave frequencies, the substrate's surface roughness increasingly becomes a factor in the insertion loss.

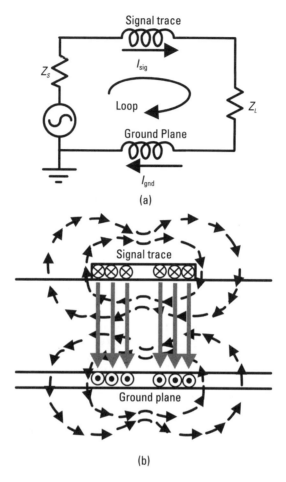

Figure 8.25 Common-mode current flow, shown (a) schematically and (b) electrically. The dotted lines denote the magnetic field and the solid lines the electric field.

8.7.2 Ground Paddle

Inside the package, the die is usually attached to a ground paddle. Similarly, the carrier often has a grounded area reserved for the package with plated-thru holes or filled vias to ground. When the attached material drains into plated-thru holes, the via's inductance will change. On the other hand, gold-plated filled vias have conductive epoxy already filled in. Yet the epoxy filler has a dielectric constant different from the substrate. The frequency of operation and resulting skin depth determines the depth of penetration of the RF current into the filler, affecting its loss.

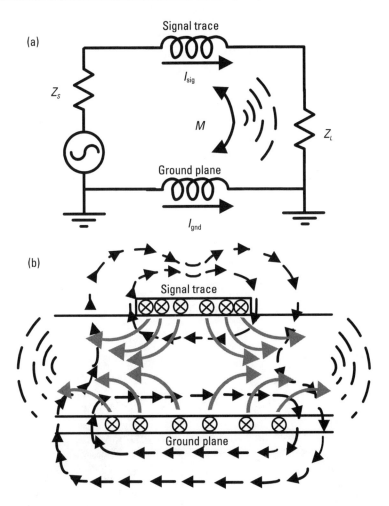

Figure 8.26 Differential mode current, shown schematically (a) and electrically (b). The dotted lines denote the magnetic field and the solid lines the electric field.

One way to measure the ground paddle's inductance is by inserting a coaxial SMA connector through the carrier and package [36]. Boring a hole through the carrier and package large enough to accept an SMA center pin, solder it to the ground paddle inside (see Figure 8.27). The connector's housing is soldered to the ground plane on back of the carrier. A simple equivalent circuit for the ground paddle is a parallel LC circuit to the ground. Measuring the one-port S-parameter S_{11} at low frequencies, the ground inductance L_g and ground capacitance C_g of the paddle are

$$L_g \approx Z_0 \frac{1 + \sec(\angle S_{11})}{2\pi f \tan(\angle S_{11})} \qquad (8.17)$$

$$C_g \approx Z_0 \frac{1}{4\pi^2 f_r^2 L_g} \qquad (8.18)$$

where f_r is the measured resonant frequency of the parallel $L_g C_g$ and Z_0 is the characteristic impedance (usually 50Ω).

As illustrated in Figures 8.25 to 8.26, the fixture's ground plane should closely couple to the signal path. To ensure ground integrity, map the ground path through the fixture as it travels in parallel with the signal path. The signal and ground paths should closely couple as the RF travels from the input launcher through the DUT to the output launcher. The ground path should be continuous, permitting image currents to flow unimpeded through the carrier, package, and die. Indirect ground paths are what lead to ground inductance and parasitic coupling.

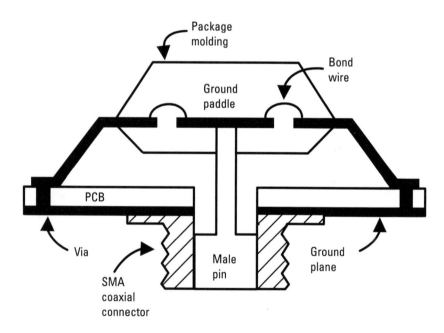

Figure 8.27 Measuring the RF characteristics of the ground paddle using a coaxial connector. The center pin extends through the PCB and the bottom of the package. The package's pins are grounded internally to the ground paddle or externally to a ground via.

8.8 Designing the Interconnecting Lines on the Carrier

RF interconnects on the carrier substrate run from the RF launchers to the package's RF ports. The choice of interconnecting transmission line (i.e., microstrip, CPW, and stripline) depends on the quantity being measured. For millimeter-wave measurements, a CPW chip can be mounted to launch the RF inside a package (see Figure 8.3). At microwave frequencies, the ground paddle inside the package can interact with the CPW to create the CBCPW mode. To avoid such issues, this section explains the choices in RF interconnects, focusing on how to apply them to package test fixture design.

8.8.1 CPW or Microstrip

While the most common choice for RF input and output lines is microstrip, in some cases CPW is preferable. To decide which to choose, a quick comparison is useful.

Nearly all RF packages require at least one ground connection. With microstrip, the ground plane is on the bottom of the substrate. To bring the ground up to the package mounted on the microstrip surface requires a ground via, yet the inductance of the via becomes prohibitive as the frequency increases. An advantage of CPW over microstrip is that ground vias are not necessary since the grounds are already on the top surface. Avoiding vias reduces ground inductance presented to the package, enhancing the performance of devices such as FETs.

In microstrip, the electric field is principally confined to the substrate; therefore, it is impacted by loss in the substrate's dielectric. Contrary to microstrip, the characteristic impedance Z_0 of a CPW line is determined by the slot width, as well as line width. To support proper wave propagation, the CPW ground-to-ground spacing should be less than a tenth of a wavelength ($\lambda/10$). Z_0 then becomes independent of frequency over a wider band.

Because of the two dielectrics (air and substrate), wave propagation in both CPW and microstrip is quasi-TEM. In CPW, the electric field penetrates the air more than in microstrip. Since air has less loss than the substrate, the line loss in CPW is less than in comparable line-width microstrip.

A drawback to CPW is that both even- and odd-current modes exist (see top of Figure 3.1 in Chapter 3). The normal CPW propagation mode is the odd mode, where the electric field vectors in the slots point in opposite directions. When the vectors point in the same direction, an even mode results, which should be suppressed (see Figure 3.2 in Chapter 3). Short-circuiting the two CPW grounds with air bridges across the signal line curtails the undesirable even mode. At the same time, it increases the cutoff frequency, leading to single-mode operation.

When a CPW substrate is mounted on a grounded carrier, it resembles CBCPW and is subject to moding (see Section 3.1 in Chapter 3). A wafer placed on a grounded chuck is a practical example of CPW transformed to CBCPW. In this instance, to keep the chuck from looking like conductor backing on the CPW, place a piece of Styrofoam between the wafer and the chuck. Styrofoam has a dielectric constant close to air ($\varepsilon_R = 1$). Using a piece as thin as 0.19 inch, the Styrofoam will decouple the chuck from the CPW, avoiding CBCPW [37].

8.8.2   CPW or CBCPW

CPW is converted to CBCPW simply by plating the backside of the substrate (see Figure 3.1 in Chapter 3). In CBCPW, three principle modes exist: the CPW mode, the parallel-plate mode, and the microstrip mode [38]. The CPW mode was discussed in Section 8.8.1. The parallel-plate mode is the result of a potential difference between the CPW ground and the backside conductor.

Depending on the assembly, CBCPW can create two distinct parallel-plate modes. One occurs between the coplanar grounds and the substrate backside metallization, resembling the two parallel plates of a capacitor. The other mode arises from the CPW grounds and the lid, assuming a conductive lid is placed over the entire assembly. Either of the two scenarios acts to guide waves. The mode due to the cover or lid is the least important since the lid is usually a long distance away from the substrate.

The principal drawback of CBCPW compared with CPW is that CBCPW leaks power into the substrate or into surface waves. Energy leakage gives rise to crosstalk. In the parallel-plate mode between the CPW grounds and backside conductor, only the lowest-order mode leaks significant energy. This occurs when the phase velocity v_p of the leaky parallel-plate mode slows to the v_p of the desired CPW mode. When the v_p's of the two modes match, they interact. This occurs at discrete frequencies. Overly wide CPW grounds can also leak energy into the substrate.

Proper design of either CPW or CBCPW can avoid energy loss to the substrate. Thin substrates, wide signal lines, and wide slots are what increase the leakage problem. One solution is to short the CPW grounds to the backside ground at the edges of the substrate. Ground vias through the substrate can perform the shorting. Vias placed intermittently through the grounds near the edges can suppress unwanted modes, improving the interconnect's insertion loss [39].

As discussed in Section 5.2 in Chapter 5, coaxial-to-CPW transitions are often found in test fixtures. A coaxial launcher can stimulate CBCPW modes such as microstrip and parallel plate [40]. Placing via holes to the backside ground intermittently in the CBCPW grounds eliminates such moding. TRL, a popular calibration method in these situations, often fails when using CBCPW because of multiple modes [41].

8.9 Quantifying the RF Effect of the Package on the Die

The presence of the package will undoubtedly affect the die inside. While specifics depend on the die and package layout, there are quantities that can guide a general understanding of their interdependence. Studying the characteristics of a transmission line, namely, its effective relative dielectric constant ε_{eff}, propagation constant γ, characteristic impedance of the line Z_0, and loss tangent $\tan\delta$, is a way to quantify these interdependencies. The transmission line is the fundamental basis for all high-frequency circuit design. This section explains how to quantify the effects of a package on a simple transmission line, leading to a generic understanding of the effects of the package on the die.

8.9.1 Effective Relative Dielectric Constant ε_{eff}

To determine how a transmission line's effective relative dielectric constant ε_{eff} is impacted by the package, mount two identical transmission lines of lengths l_1 and l_2 inside the package. With the VNA, measure the phase difference $(\theta_2 - \theta_1)$ between the two transmission lines, applying the results to the following equation

$$\varepsilon_{eff} = \left(\frac{c(\theta_2 - \theta_1)}{360 f(l_2 - l_1)} \right)^2 = \left(\frac{c}{(\lambda_2 - \lambda_1) f} \right)^2 \tag{8.19}$$

where c is the speed of light (m/s), λ_1 and λ_2 are the wavelengths of the two lines (m), f is the measurement frequency (in hertz), l_1, l_2 are the physical lengths of the lines (m), and θ_1, θ_2 are their electrical lengths (degrees). This reveals the change in a transmission line's ε_{eff} due to the presence of the package.

8.9.2 Propagation Constant γ and Characteristic Impedance Z_0

The propagation constant γ and characteristic impedance Z_0 of a transmission line can be found from the two-port S-parameters of a packaged transmission line. Inserting the measured S-parameters into the following equation [42]

$$e^{-\gamma l} = \left[\frac{1 - S_{11}^2 + S_{21}^2}{2S_{21}} \pm K \right]^{-1} \tag{8.20}$$

where l is the line length (in meters). Only one root of (8.20) is realizable. Also

$$K = \left[\frac{\left(S_{11}^2 - S_{21}^2 + 1 \right)^2 - \left(2S_{11} \right)^2}{\left(2S_{21} \right)^2} \right]^{\frac{1}{2}} \tag{8.21}$$

$$Z_0^2 = Z_{\text{ref}}^2 \frac{\left(1 + S_{11} \right)^2 - S_{21}^2}{\left(1 - S_{11} \right)^2 - S_{21}^2} = \frac{R + j\omega L}{G + j\omega C} \tag{8.22}$$

The reference impedance Z_{ref} is typically 50Ω. Use (8.21) and (8.22) to solve (8.19) for γ. Knowing γ and Z_0, the series impedance ($Z_s = R + jX$) and shunt admittance ($Y_p = G + jC$) of a transmission line can be found by

$$R = \text{Re}[\gamma Z_0] \tag{8.23}$$

$$L = \frac{\text{Im}[\gamma Z_0]}{\omega} \tag{8.24}$$

$$G = \text{Re}\left[\frac{\gamma}{Z_0} \right] \tag{8.25}$$

$$C = \frac{\text{Im}\left[\frac{\gamma}{Z_0} \right]}{\omega} \tag{8.26}$$

For a CPW line [43]

$$G + j\omega C = \frac{\gamma}{Z_0} \tag{8.27}$$

where G is the conductance and C is the capacitance of a CPW transmission line. Note that the propagation constant of a general transmission line is classically defined as

$$\gamma = \sqrt{Z_s Y_p} = \sqrt{(R + j\omega L)(G + j\omega C)} \qquad (8.28)$$

Z_s is the series impedance and Y_p is the shunt admittance of the CPW line.

In an alternate approach, the propagation constant γ can be calculated during a TRL calibration [44, 45] and the line's capacitance C measured at low frequency. In insulating GaAs, G is negligible [46]. By neglecting G and knowing γ and C, Z_0 can then be found using (8.27).

8.9.3 Loss Tangent tanδ

To find the loss tangent tanδ of a packaged transmission line, a ring resonator is useful. The circumference L of the circular ring is defined by [47]

$$L = n\lambda_g \qquad (8.29)$$

where λ_g is the lowest-order resonance of the ring and n the mode number. Since

$$\lambda_g = \frac{c}{f\sqrt{\varepsilon_{\text{eff}}}} \qquad (8.30)$$

Inserting (8.30) into (8.29) gives the effective relative dielectric constant ε_{eff} due to the presence of the package

$$\varepsilon_{\text{eff}} = \left[\frac{nc}{fL}\right]^2 \qquad (8.31)$$

where c is the speed of light (m/s) and f the ring's resonant frequency (in hertz). The quality factor Q of a resonator is given by

$$Q = \frac{f}{BW_{3\text{dB}}} \qquad (8.32)$$

where $BW_{3\text{dB}}$ is the 3-dB bandwidth at its resonant frequency. The Q of the resonator is the sum of three components, the conductor Q_{cond}, the radiated Q_{rad}, and the dielectric Q_{diel} components

$$\frac{1}{Q} = \frac{1}{Q_{\text{cond}}} + \frac{1}{Q_{\text{rad}}} + \frac{1}{Q_{\text{diel}}} \qquad (8.33)$$

Q_{cond} and Q_{rad} can be found by calculation or electromagnetic simulation, leaving Q_{diel} as the measured part. The loss tangent is [48]

$$\tan \delta = \frac{\varepsilon_{eff}(\varepsilon_r - 1)}{Q_{diel}\varepsilon_r(\varepsilon_{eff} - 1)} \qquad (8.34)$$

Note that at frequencies below 5 GHz, the ring can be physically large. Also, CPW resonators are available [49].

8.10 Package Styles

This section discusses the RF characterization of particular package styles, focusing on those that are popular today, as well as those newly arriving on the market.

8.10.1 Plastic Surface Mount Packages

The plastic surface mount package is the most common package style on the market today. The die, bond wires, and lead frame are all encapsulated in a black molding compound. The task is to accurately characterize the effect of the molding compound on the component's RF performance.

As discussed in Section 8.9, the impact of the dielectric compound on the die can be quantified by four fundamental quantities: the effective relative dielectric constant ε_{eff}, the propagation constant γ, the characteristic impedance Z_0, and the loss tangent $\tan\delta$. The best characterization techniques are the transmission line technique and resonant technique [50]. The transmission line technique encapsulates a length of transmission line to find γ and Z_0 (see Sections 8.9.1 to 8.9.2). The resonant technique uses the resonance of an encapsulated resonator to find its quality factor Q, leading one to find ε_{eff} and $\tan\delta$ (see Section 8.9.3).

8.10.2 Flip-Chip

Flip-chip offers a reduction in package size while at the same time shortening the lengths of the RF interconnects within the package. The simplest form of flip-chip is a bare die with gold/nickel (Au/Ni) ball bumps deposited on the die's bond pads. Flip-chipped die are often designed to land on CPW transmission lines since CPW provides grounding on the top surface. Flip-chip bumps have less parasitic inductance and capacitance than conventional package interconnects (see Figure 8.28). The bump design is optimized to reduce losses and reflections at the interface.

Flipping the chip affects the die's RF performance in two ways. In the first way, the characteristic impedance Z_0 of the CPW line changes due to the proximity of the substrate underneath (see Figure 8.29). To avoid this interaction, the minimum gap h should be no less than the distance g between grounds of the CPW line [51]. The bump height h should be as short as possible [52, 53]. The size of the bump pad also increases the capacitance. The overlap o of the CPW lines results in fringing capacitance between the die and carrier. This overlap is governed by chip dicing and automated chip placement process tolerances.

Critical factors to review when designing the bumps include the following:

- The overlap o between the die and carrier conductors;
- The bump's diameter;
- The width w of the CPW signal line;
- The pad size on the substrate.

In general, the bump-to-substrate transition is capacitive, although at higher frequencies it becomes inductive due to the skin effect of the bump. The larger the bump, the lower the frequency at which its inductance will dominate. To measure the S-parameters of the transition alone, TDR can be

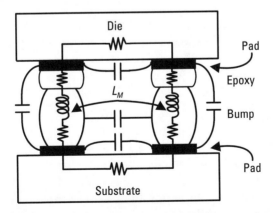

Figure 8.28 Equivalent circuit of a signal and ground bump of a flip-chip die on a carrier substrate. Inductive cross-coupling L_M can also occur between bumps.

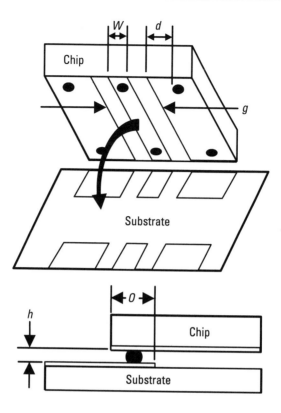

Figure 8.29 Flip-chip mounted to a substrate (no underfill).

used to window around the bump [54]. When there is no backside ground on the die or substrate, the die and substrate thicknesses will not be as important. Otherwise, with CBCPW, moding to the carrier substrate is possible. Modes are commonly launched at discontinuities such as the bump-to-substrate transition [55, 56]. To investigate moding, measure a CPW short on a die that is flip-chipped onto the carrier (see middle of Figure 3.17 in Chapter 3).

The capacitance of the bumps can be compensated for by increased inductance on either the die or substrate. Widening the space between the two CPW ground bumps results in a longer ground path and more inductance. A more complicated solution to excessive capacitance is to design a tuning network on the carrier substrate at the bump pad.

Staggering the alignment of the bumps will decrease their capacitance. Ensure the bumps are not in-line with one another by setting the two ground bumps back from the signal bump. When doing so, take care to define the reference plane properly during calibration.

As the die nears the surface of the substrate, it alters the electrical performance of the die, although this diminishes for larger bump heights. To detect a change in RF behavior due to proximity of the carrier substrate, lay out a simple thru on the die and observe how its phase varies from linear [57]. An alternate method is to look for a change in the resonant frequency of a resonator, where the substrate changes its effective relative dielectric constant ε_{eff} (see Section 8.3.2).

To reduce stress on the bumps, underfill is often applied between the die and substrate. Because the dielectric constant of the underfill differs from air, the underfill perturbs the die similar to the presence of the substrate [58]. To reduce the effect of die coupling through the underfill, minimize the height of the bump.

Keeping the test fixture clean avoids particles from getting lodged between the bumps of the flip-chip. When mounting the flip-chip package onto the carrier substrate, excessive pressure will compress the bumps and lessen the distance between the package and substrate, altering the bump's impedance.

Ball-grid array (BGA) is similar to flip-chip in that both use ball bumps, hence, many of the same bump issues arise. BGA is a bare die flipped onto a layer of either thick-film or PCB [59]. The PCB acts as a thin interconnect layer that redistributes the die pad connections to a wider pitch, allowing a wider spacing on the carrier's pad layout.

8.10.3 Bumped Chip Carrier

Also known as *metal lead frame* (MLF), the *bump chip carrier* (BCC) package is smaller than the *small outline transistor* (SOT) package, a comparable assembly. The BCC has a metal paddle in the center to hold the die and is encircled by thick metal bond pads. The bond pads and ground paddle provide a low inductance path to the leadless external pads. Wire bonds connect the die in the center to the bond pads around its periphery. The completed BCC assembly is encapsulated with a resin overmolding for protection.

Prior to encapsulation, the empty BCC package can be RF-characterized using coplanar probes [60]. The drawback is that G-S probe pitch must match the package's pad-to-ground paddle spacing (see Figure 8.30). Because RF probing is done before the resin overmolding is applied, the die's RF behavior will be different from its final packaged performance.

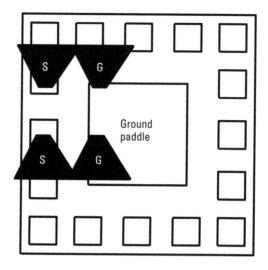

Figure 8.30 Using coplanar probes to characterize the leads of a BCC++16 package.

8.11 Summary

As the market trend progresses toward smaller package sizes (less than 4 mm²) and closer pad spacing (less than 0.5 mm), package characterization will become even more difficult. At the same time, the drive toward lower cost makes fast and accurate package characterization essential to a design's first-pass success.

As the industry moves toward higher levels of component integration, the package will become more complex and require a greater number of tests. Multiple die in the same package will mean that interaction between die will go beyond the simple mutual inductance and capacitance occurring between bond wires and leads in a package [61].

References

[1] Chun, C., et al. "Development of Microwave Package Models Utilizing On-Wafer Characterization Techniques," *IEEE Trans. on Microwave Theory and Techniques*, Vol. 45, No. 10, 1997, pp. 1948–1954.

[2] Smith, S., and M. Murphy, "Electrical Characterization of Packages for Use with GaAs MMIC Amplifiers," *IEEE Microwave Symposium Digest*, 1993, pp. 131–134.

[3] J Micro Technology, "Test Adapter Substrates for PHEMT FET Measurements," *Microwave Journal*, Vol. 38, No. 3, 1995, pp. 120–122.

[4] Monahan, G., A. Morris, and M. Steer, "A Coaxial Test Fixture for Characterizing Low-Impedance Microwave Two-Terminal Devices," *Microwave and Optical Technology Letters*, Vol. 6, No. 3, 1993, pp. 197–200.

[5] Kazi, K., B. Szendrenyi, and I. Mojzes, "Lossy Model of Diode Packages: An Alternative Method for Exact Evaluation of Active Chip Parameters," *IEEE Microwave Symposium Digest*, 1989, pp. 1267–1270.

[6] Bastos, J., "Influence of Die Attachment on MOS Transistor Matching," *IEEE International Conference on Microelectronic Test Structures*, Vol. 9, 1996, pp. 27–31.

[7] Ponchak, G., et al. "The Use of Metal Filled Via Holes for Improving Isolation in LTCC RF and Wireless Multichip Packages," *IEEE Trans. on Advanced Packaging*, Vol. 23, No. 1, 2000, pp. 88–99.

[8] Tarvainen, T., J. Kolehmainen, and E. Oy, "On 2D Transmission Line Modeling of Power and Ground Planes in Multilayer Structures," *IEEE Conference on Electrical Performance of Electronic Packaging*, 2000, pp. 95–98.

[9] Mouthaan, K., et al., "Microwave Modeling and Measurement of the Self and Mutual Inductance of Coupled Bond Wires," *IEEE Bipolar Circuits and Technology Meeting*, 1997, pp. 166–169.

[10] Nelson, S., et al., "Optimum Microstrip Interconnects," *IEEE Microwave Symposium Digest*, 1991, pp. 1071–1074.

[11] Budka, T., "Wide-Bandwidth Millimeter-Wave Bond-Wire Interconnects," *IEEE Trans. on Microwave Theory and Techniques*, Vol. 49, No. 4, 2001, pp. 715–718.

[12] Qi, X., et al., "A Fast 3D Modeling Approach to Electrical Parameters Extraction of Bonding Wires for RF Circuits," *IEEE Trans. on Advanced Packaging*, Vol. 23, No. 3, 2000, pp. 480–488.

[13] Caggiano, M., et al., "Electrical Modeling of the Chip Scale BGA," *IEEE Electronic Components and Technology Conference*, 1998, pp. 1280–1285.

[14] Dernevik, M., et al., "Electrically Conductive Adhesives at Microwave Frequencies," *IEEE Electronic Components and Technology Conference*, 1998, pp. 1026–1030.

[15] Sihlbom, R., et al., "Conductive Adhesives for High-Frequency Applications," *IEEE Trans. on Components, Packaging, and Manufacturing Technology—Part A*, Vol. 21, No. 3, 1998, pp. 469–477.

[16] Felba, J., K. Friedel, and A. Moscicki, "Characterization and Performance of Electrically Conductive Adhesives for Microwave Applications," *IEEE Electronic Components and Technology Conference*, 2000, pp. 232–239.

[17] Godshalk, E., "Characterization of Surface Mount Packages at Microwave Frequencies Using Wafer Probes," *IEEE Microwave Symposium Digest*, 2000, pp. 1887–1890.

[18] Jackson, R., "A Circuit Topology for Microwave Modeling of Plastic Surface Mount Packages," *IEEE Trans. on Microwave Theory and Techniques*, Vol. 44, No. 7, 1996, pp. 1140–1146.

[19] Jackson, R., and S. Rakshit, "Microwave-Circuit Modeling of High Lead-Count Plastic Packages," *IEEE Trans. on Microwave Theory and Techniques*, Vol. 45, No. 10, 1997, pp. 1926–1933.

[20] Jackson, R., "Modeling Millimeter-Wave IC Behavior for Flipped-Chip Mounting Schemes," *IEEE Trans. on Microwave Theory and Techniques*, Vol. 45, No. 10, 1997, pp. 1919–1925.

[21] Hauwermeiren, L., M. Botte, and D. De Zutter, "A New De-embedding Technique for On-Board Structures Applied to the Bandwidth Measurement of Packages," *IEEE Trans. on Components, Hybrids, and Manufacturing Technology*, Vol. 16, No. 3, 1993, pp. 300–303.

[22] Martin, A., and M. Dukeman, "Measurement of Device Parameters Using a Symmetric Fixture," *Microwave Journal*, Vol. 30, No. 5, 1987, pp. 299–306.

[23] Winkel, T., L. Dutta, and H. Grabinski, "An On-Wafer De-embedding Procedure for Devices Under Measurement with Error-Networks Containing Arbitrary Line Lengths," *47th Automatic RF Techniques Group Conference Digest*, 1996, pp. 102–111.

[24] Hewlett-Packard, "Time Domain Reflectometry Theory," *Hewlett-Packard Application Note 1304-2*.

[25] Nicolson, A., "Broadband Microwave Transmission Characteristics from a Single Measurement of the Transient Response," *IEEE Trans. on Instrumentation and Measurement*, Vol. 17, No. 6, 1968, pp. 395–402.

[26] Gronau, G., and I. Wolff, "A Simple Broadband Device De-embedding Method Using an Automatic Network Analyzer with Time-Domain Option," *IEEE Trans. on Microwave Theory and Techniques*, Vol. 37, No. 3, 1989, pp. 479–483.

[27] Kompa, G., M. Schlechtweg, and F. Van Raay, "Precisely Calibrated Coaxial-to-Microstrip Transitions Yield Improved Performance in GaAs FET Characterization," *IEEE Trans. on Microwave Theory and Techniques*, Vol. 38, No. 1, 1990, pp. 62–68.

[28] Ferrero, A., F. Sanpetro, and U. Pisani, "Multiport Vector Network Analyzer Calibration: A General Formulation," *IEEE Trans. on Microwave Theory and Techniques*, Vol. 42, No. 12, 1994, pp. 2455–2461.

[29] Tsai, C.-T., and W.-Y. Yip, "An Experimental Technique for Full Package Inductance Matrix Characterization," *IEEE Transactions on Components, Packaging, and Manufacturing Technology – Part B*, Vol. 19, No. 2, 1996, pp. 338–343.

[30] Young, B., and A. Sparkman, "Measurement of Package Inductance and Capacitance Matrices," *IEEE Trans. on Components, Packaging, and Manufacturing Technology—Part B*, Vol. 19, No. 1, 1996, pp. 225–229.

[31] Caggiano, M., S. Bulumulla, and D. Lischner, "RF Electrical Measurements of Fine Pitch BGA Packages," *IEEE Electronic Components and Technology Conference*, 2000, pp. 449–453.

[32] Hietala, V., "Determining Two-Port S-Parameters from a One-Port Measurement Using a Novel Impedance-State Test Chip," *IEEE Microwave Symposium Digest*, 1999, pp. 1639–1642.

[33] Sercu, S., and L. Martens, "High-Frequency Circuit Modeling of Large Pin Count Packages," *IEEE Trans. on Microwave Theory and Techniques*, Vol. 45, No. 10, 1997, pp. 1897–1904.

[34] Deng, J., and H.-K. Chiou, "Considerations of Characterizing Standard SMT Packages for RFIC Applications," *IEEE Conference on Electrical Performance of Electronic Packaging*, 1998, pp. 97–100.

[35] Young, B., and A. Sparkman, "Measurement of Package Inductance and Capacitance Matrices," *IEEE Trans. on Components, Packaging, and Manufacturing Technology—Part B*, Vol. 19, No. 1, 1996, pp. 225–229.

[36] Horng, T., S. Wu, and C. Shih, "Electrical Modeling of RFIC Packages up to 12 GHz," *IEEE Electronic Components and Technology Conference*, 1999, pp. 867–872.

[37] Hammond, C., and K. Virga, "Network Analyzer Calibration Methods for High-Density Packaging Characterization and Validation of Simulation Models," *IEEE Electronic Components and Technology Conference*, 2000, pp. 519–525.

[38] Hasegawa, H., M. Furukawa, and H. Yanai, "Properties of Microstrip Line on Si-SiO$_2$ System," *IEEE Trans. on Microwave Theory and Techniques*, Vol. 19, No. 11, 1971, pp. 869–881.

[39] Tien, C., et al., "Transmission Characteristics of Finite-Width Conductor-Backed Coplanar Waveguide," *IEEE Trans. on Microwave Theory and Techniques*, Vol. 41, No. 9, 1993, pp. 1616–1624.

[40] Yu, M., R. Vahldieck, and J. Huang, "Comparing Coax Launcher and Wafer Probe Excitation for 10-mil Conductor Backed CPW with Via Holes and Airbridges," *IEEE Microwave Symposium Digest*, 1993, pp. 705–708.

[41] Williams, D., J. Belquin, and G. Dambrine, "On-Wafer Measurement at Millimeter-Wave Frequencies," *IEEE Microwave Symposium Digest*, 1996, pp. 1683–1686.

[42] Eo, Y., and W. Eisenstadt, "High-Speed VLSI Interconnect Modeling Based on S-Parameter Measurements," *IEEE Trans. on Components, Hybrids, and Manufacturing Technology*, Vol. 16, No. 6, 1993, pp. 555–562.

[43] Marks, R., and D. Williams, "Characteristic Impedance Determination Using Propagation Constant Measurement," *IEEE Microwave and Guided Wave Letters*, Vol. 1, 1991, pp. 141–143.

[44] Marks, R., "A Multiline Method of Network Analyzer Calibration," *IEEE Trans. on Microwave Theory and Techniques*, Vol. 39, No. 7, 1991, pp. 1205–1215.

[45] Janezic, M., and D. Williams, "Permittivity Characterization from Transmission-Line Measurement," *IEEE Microwave Symposium Digest*, 1997, pp. 1343–1346.

[46] Williams, D., and R. Marks, "Transmission Line Capacitance Measurement," *IEEE Microwave and Guided Wave Letters*, Vol. 1, No. 9, 1991, pp. 243–245.

[47] Gipprich, J., et al., "Microwave Dielectric Constant of a Low Temperature Cofired Ceramic," *IEEE Trans. on Components, Hybrids, and Manufacturing Technology*, Vol. 14, 1991, pp. 732–737.

[48] Amey, D., and J. Curilla, "Microwave Properties of Ceramic Materials," *IEEE Electronic Components and Technology Conference*, 1991, pp. 267–272.

[49] Waldo, M., I. Kaufman, and S. El-Ghazaly, "Coplanar Waveguide Technique for Measurement of Dielectric Constant or Thickness of Dielectric Films," *IEEE Microwave Symposium Digest*, 1997, pp. 1339–1342.

[50] Riad, S., et al., "Wideband Electrical Characterization and Modeling of Plastic Packaging Materials," *IEEE International Symposium on Advanced Packaging Materials*, 1998, pp. 129.

[51] Wong, Y., L. Pattison, and D. Linton, "Flip-Chip Interconnect Analysis at Millimeter-Wave Frequencies," *IEEE High Frequency Postgraduate Student Colloquium*, 1999, pp. 82–87.

[52] Staiculescu, D., J. Laskar, and E. Tentzeris, "Design Rule Development for Microwave Flip-Chip Applications," *IEEE Trans. on Microwave Theory and Techniques*, Vol. 48, No. 9, 2000, pp. 1476–1481.

[53] Kusamitsu, H., et al., "The Flip-Chip Bump Interconnection for Millimeter-Wave GaAs MMIC," *IEEE Trans. on Electronics Packaging Manufacturing*, Vol. 22, No. 1, 1999, pp. 23–28.

[54] Wong, A., and D. Linton, "Copper Flip Chip Bump Interconnect Technology for Microwave Subsystems Including RF Characterization," *IEEE Electronics Packaging Technology Conference*, 2000, pp. 335–338.

[55] Jackson, R., and R. Ito, "Modeling Millimeter-Wave IC Behavior for Flipped-Chip Mounting Schemes," *IEEE Trans. on Microwave Theory and Techniques*, Vol. 45, No. 10, 1997, pp. 1919–1925.

[56] Jackson, R., "Mode Conversion at Discontinuities in Finite-Width Conductor-Backed Coplanar Waveguide," *IEEE Trans. on Microwave Theory and Techniques*, Vol. 37, No. 10, 1989, pp. 1582–1588.

[57] Jentzsch, A., and W. Heinrich, "Theory and Measurements of Flip-Chip Interconnects for Frequencies Up to 100 GHz," *IEEE Trans. on Microwave Theory and Techniques*, Vol. 49, No. 5, 2001, pp. 871–878.

[58] Feng, Z., et al., "RF and Mechanical Characterization of Flip-Chip Interconnects in CPW Circuits with Underfill," *IEEE Trans. on Microwave Theory and Techniques*, Vol. 46, No. 12, 1998, pp. 2269–2275.

[59] Caggiano, M., et al., "RF Electrical Measurements of Fine Pitch BGA Packages," *IEEE Trans. on Components and Packaging Technologies*, Vol. 24, No. 2, 2001, pp. 233–240.

[60] Horng, T., et al., "Electrical Performance Improvements on RFICs Using Bump Chip Carrier Packages as Compared to Standard Small Outline Packages," *IEEE Electronic Components and Technology Conference*, 2000, pp. 439–444.

[61] Gharpurey, R., and R. Meyer, "Modeling and Analysis of Substrate Coupling in Integrated Circuits," *IEEE Journal of Solid-State Circuits*, Vol. 31, No. 3, 1996, pp. 344–353.

9

Future Trends

When wireless products reach their potential as affordable, mobile, and seamless personal communications devices, then the RFICs and MMICs inside will have become a part of everyday life. The Bluetooth and IEEE 802.11b standards are examples of recent attempts to expand the influence of RF components. Both are 2.4-GHz RF link protocols for communicating with common consumer appliances [1]. While optical networks are available to replace some RF links, the low data rates of the average household appliance do not require the blinding speed of optical interfaces, which make more sense for large-capacity trunk lines.

The first part of this chapter gives an overview of how RF testing fits into today's IC design cycle. The rest of the chapter discusses future trends—primarily driven by wireless products—in RF testing.

9.1 The Typical Design Cycle

The amount of thought the RF circuit designer spends at the start of a project can have a tremendous effect on the outcome. Beginning a project is analogous to launching a rocket. A few feet off-course as it leaves the launch pad will mean hundreds of miles off-course by the time it reaches outer space. A successful RF design requires a solid foundation.

For a successful product, the IC designer has numerous areas to contend with, including the following [2]:

- IC fabrication;
- Packaging;
- Assembly;
- Reliability;
- Testing.

How well these are balanced often determines the project's success. The RF product designer typically works with a specialist in each of the areas mentioned. The IC fabrication process can be complicated, often involving dozens of mask layers and hundreds of deposition, etching, and masking steps. What starts as a bare wafer ends in a functioning IC.

Consider the package surrounding the IC. The designer has to understand how the package's RF characteristics affect the IC's RF performance. There are assembly issues, such as how the die is mounted into the package. It can be soldered or epoxied, either conductively or nonconductively. A number of other questions arise. How do the signal and ground currents flow through the package to the die? How does the part hold up under environmental conditions and over time (its reliability)? Are the CTEs of the materials vastly different? Furthermore, is the IC designed to permit testing without an overcomplicated setup? Of the issues listed, the one this book deals with is RF testing.

Once the domain of RF analog circuit designers, on-wafer probing has entered the world of high-speed digital circuits. Recently, microprocessors have crossed the 1-GHz clock speed threshold. Gigabits per second of data are traveling along high-speed Internet trunk lines. Because square-wave digital data pulses create harmonics into the microwave and millimeter-wave range, accurate probing of high-speed digital ICs requires the use of RF techniques, using either coplanar probes or test fixtures.

9.2 The Separate Worlds of Digital and RF

Most handheld wireless devices consist of two fundamental sections, the RF transceiver section and the *digital-signal processing* (DSP) section. In the RF section, an RF signal from the antenna is mixed (or downconverted) to an IF signal. A digital chip then performs DSP operations such as de-encoding and signal interference reduction on the IF data stream.

Most standard DSP chips are based on silicon bipolar complementary metal-oxide semiconductor (BiCMOS) FET technology. The RF section can be composed of discrete chips made from different technologies, such as Si, GaAs, or silicon germanium (SiGe). Each of these fabrication technologies has its own advantages and disadvantages. For instance, components such as capacitors and inductors are not easily brought onto an Si RF chip [3]. In general, RFIC chip technology has not benefited from many of the advances in digital IC technology.

Unlike digital design, designing an RF component cannot be done solely by *computer-aided design* (CAD). It requires an understanding of a variety of RF disciplines, such as electromagnetic theory, communications theory, modulation techniques, and transceiver system architecture. Any one of these can require years of study to grasp. Hence, the RF designer often relies on experience and intuition rather than CAD to predict performance. To gain a hint at what is involved, the fundamental tradeoffs in an RF design are summed up in Figure 9.1. Arrows are drawn between the related elements. For instance, the supply voltage directly affects both the linearity of an amplifier and its RF gain.

9.3 The Goal: The Marriage of Digital and RF into a Single Wireless Product

Microelectronics is constantly pushing the boundaries of what is achievable on a single chip. The near-term vision for wireless products is to integrate both the analog and digital functions onto the same chip [4]. At the moment, bringing the separate analog and digital CAD tools together to analyze the

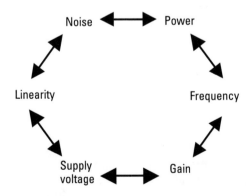

Figure 9.1 Six important tradeoffs in an RF design.

complete chip is a complicated task. Yet once achieved, the handset will be free to evolve into a wireless data terminal that can handle voice, data, and video along with on-board computing [5].

One of the most severe constraints on an RF design is the limited spectrum available. The digital section employs encoding, compression, and modulation schemes requiring a high bit rate. High bit rates use a large RF bandwidth. In the RF section, the transmitter must be narrowband enough to avoid interfering with adjacent channels. The RF receiver must reject any out-of-band interference. In general, the receiver's dynamic range is limited on the low end by the noise floor and on the upper end by nonlinearities and saturation.

The ultimate dream is to completely eliminate the analog by digitizing the signal received from the antenna, thereafter operating solely in the digital domain. The RF signal received from the antenna would be immediately digitized, filtered, and sent to low-frequency DSP. Per the Nyquist principle, the rate of the *analog-to-digital conversion* (ADC) must be at least twice the frequency of the RF signal (greater than 4 GHz) and have a dynamic range of greater than 100 dB. Employing such an A/D converter would push the digital interface closer to the antenna.

The heterodyne principle in RF theory rests on the conversion (or mixing) of the RF to a lower IF. The IF is usually further downconverted to a baseband signal. If the RF signal itself could be digitized as it is received, then many of the components in the receiver chain (particularly the off-chip discrete components) could be eliminated. Direct downconversion (also known as a homodyne conversion) converts a signal from RF to baseband in one mixing step. By mixing the RF signal with a single LO, the resulting signal can be lowpass filtered and centered around zero frequency. Because the high-frequency noise will also be downconverted, a low-noise amplifier often precedes the mixer. There are a number of challenges to this scheme, in particular, noise immunity. For instance, leakage between the LO and RF in the receive path must be minimal. In the transmit path, the strength of the signal coming from the PA can disturb the LO. Noise is generated on the chip due to large voltage swings.

9.4 New Substrate Materials

The system designer has little time to evaluate the multitude of choices available to designing a wireless component. Choosing the material to build the ICs on is made early in the wireless component design process. With any substrate selection, cost, power, and speed are the primary drivers, followed by packaging density and complexity. GaAs has traditionally been the material of choice for high-frequency circuits, yet others such as SiGe, indium phosphide (InP), and high-speed Si BiCMOS have recently emerged. The current problem is in integrating the multitude of technologies onto a single platform.

The high-speed capability of SiGe offers an integrated digital/RF solution, differing only slightly from a standard Si process. An all-silicon-based digital/RF solution would eliminate the assembly costs involved in multiple chips. One of the drawbacks has been the high transmission line loss on silicon-based substrates. The collector-emitter breakdown voltage (BV_{CEO}) is lower with SiGe transistors compared with Si, limiting its power-handling capability [6]. For power in handheld devices, GaAs power amplifiers [such as *heterojunction bipolar transistors* (HBTs) and *metal semiconductor field-effect transistors* (MESFETs)] still dominate. For high-power base stations, silicon carbide (SiC) and gallium nitride (GaN) show promise. Extracting the resistive and capacitive aspects of new substrates and incorporating them into a CAD tool to accurately simulate them is the challenge.

9.5   The Direction of RF Development

Prior to 1990, the RF and microwave market centered on the military, where high-volume manufacturing numbered in the thousands of units. By comparison, today's wireless market explosion pushed RF testing to a massive scale. For instance, PAs were initially designed from discrete silicon transistors and matching components located on a PCB [7]. To reduce matching sensitivity above 1 GHz, MMIC incorporated the tuning on the chip.

However, including the matching elements on a GaAs chip is more expensive than assembling a discrete transistor and discrete matching elements on a separate PCB. In a PCB component, assembling the board with *known good die* (KGD) is paramount. Throwing away an assembled board is more expensive than RF probing for good die beforehand. Cost reduction and size reduction pressures are transforming discrete transistor amplifiers into full front-end RF modules with the ultimate goal of a transceiver on a single chip.

In the future, RF circuits will develop in two directions simultaneously. One is increased functionality, adding new user features to the component. The other is stricter RF performance, calling for a higher quality of

RF testing. Developing an RF test to tackle these two simultaneously will be the challenge.

In recent years, cellular service providers have concentrated on signing up customers rather than offering more and better cell phone technology. Once the number of users levels off, the latter will catch up. Much of the delay has been in defining universal standards and their adoption by the industry, rather than identifying shortcomings in the technology.

9.6 The Future of the RF Test

In the past, RF testing was the last step in the assembly process, performed on a sample basis. From this, RF testing has evolved into an integral part of a cost-efficient production process. As complexity of the RF component increased, the need for early screening has become more important. In the future, *chip-scale packaging* (CSP) will mean that the entire wafer is packaged during the IC fabrication process [8]. To determine the yield early, the wafer will need to be RF-tested inside the foundry, with good wafers sorted and then die singulated.

In general, RF measurement accuracy is essential to a successful design. RF test accuracy does the following:

- Increases the throughput (by reducing guard bands around the specified performance);

- Improves the final delivered RF performance;

- Increases the production yield;

- Produces better component consistency;

- Better fits the CAD models.

In the IC foundry, semiconductor process variations can be costly. Spot-checking wafers with accurate RF measurements both verifies the die's performance and determines the wafer's yield. Wafer-mapping the RF parameters allows process variations to be plotted across the wafer and better understood. Wafer mapping can also speed the development of new fabrication processes [9]. Compared with bonding and fixturing individual die, wafer probing is undoubtedly much easier (see Figure 9.2).

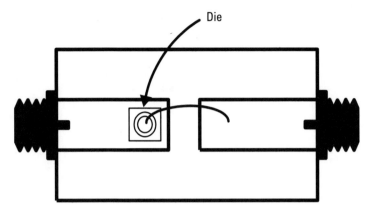

Die

Figure 9.2 Mounting an individual die to a connectorized board is the simplest way to find out the RF performance of a wafer. A vertical diode die is shown.

9.7 Summary

Although the digital portion of a wireless handset may have millions of transistors, the RF section is arguably the biggest challenge in a wireless design. The often-cited problem is the lack of CAD tools to accurately predict the RFIC's performance [10]. When a performance problem arises, the finger quickly points to the CAD model, drawing into question its theoretical basis. Yet many CAD shortcomings can just as easily be due to the RF measurements used to generate the models than to the models themselves. Employing the methods in this book will enable the reader to make better RF measurements, leading to better RFICs and MMICs in today's marketplace.

References

[1]   Meng, T., and B. McFarland, "Wireless LAN Revolution: From Silicon to Systems," *IEEE Radio Frequency Integrated Circuits Symposium Digest*, 2001, pp. 3–6.

[2]   Hashemi, H., et al., "Design for Manufacturing of RF MCM-L," *IEEE Conference on Electrical Performance of Electronic Packaging*, 2000, pp. 283–286.

[3]   Campbell, S., and A. Gopinath, "Possibility of Silicon Monolithic Millimeter-Wave Integrated Circuits," *IEEE Microwave Symposium Digest*, 1989, pp. 817–819.

[4]   Long, J., et al., "RF Analog and Digital Circuits in SiGe Technology," *IEEE International Solid-State Circuits Conference*, 1996, pp. 82–83, 423.

[5]   Razavi, B., "Challenges in Portable RF Transceiver Design," *IEEE Circuits and Devices Magazine*, Vol. 12, No. 5, 1996, pp. 12–25.

[6] Larson, L., et al., "Si/SiGe HBT Technology for Low-Cost Monolithic Microwave Integrated Circuits," *IEEE International Solid-State Circuits Conference*, 1996, pp. 80–81, 422.

[7] Jos, R., "Future Developments and Technology Options in Cellular Phone Power Amplifiers: From Power Amplifier to Integrated RF Front-End Module," *IEEE Proceedings of Bipolar/BiCMOS Circuits and Technology Meeting*, 2000, pp. 118–125.

[8] Goldstein, H., "Packages Go Vertical," *IEEE Spectrum*, Vol. 38, No. 8, 2001, pp. 46–51.

[9] Blood, W., et al., "Library Development Process for Embedded Capacitors in LTCC," *IEEE Conference on Electrical Performance of Electronic Packaging*, 2000, pp. 147–150.

[10] Mayaram, K., et al., "Computer-Aided Circuit Analysis Tools for RFIC Simulation: Algorithms, Features, and Limitations," *IEEE Trans. on Circuits and Systems—II: Analog and Digital Signal Processing*, Vol. 47, No. 4, 2000, pp. 274–286.

About the Author

Scott A. Wartenberg began his RF career as a teenager in his hometown of Southaven, Mississippi, repairing televisions and radios on the bench. In 1986, he received a B.S. in electrical engineering with honors from the University of Tennessee in Knoxville. He received an M.S. in 1991 and a Ph.D. in 1997, both in electrical engineering, from the Johns Hopkins University in Baltimore, Maryland. Beginning in 1986 with the U.S. Department of Defense in Washington, D.C., his RF/microwave career has spanned microwave multichip module (MCM) and MMIC design for radar applications as well as more general RF on-wafer and module test techniques for Raytheon, Westinghouse, and Agilent Technologies. He is currently a staff engineer with RF Micro Devices in Greensboro, North Carolina. A senior member of the IEEE and a member of the Microwave Theory and Techniques Society, Dr. Wartenberg has published 20 technical articles.

Index

Absorber, radio frequency, 60
ACA. *See* Anistropic electrically conductive adhesive
Accuracy, 48
Active components, test fixture, 99
Active on-wafer device, 115, 119
Adapter removal, 60–61
Adapters. *See* Error adapter; Radio frequency connector adapters
ADC. *See* Analog-to-digital conversion
Adhesives, conductive, 168–70
Admittance noise correlation matrix, 132–33
Air, pads suspended in, 77–78, 127
Air bridge, 138–39
Alignment, 11
  coplanar probe, 63–64
  membrane probe, 91–92
Alumina, 65, 67, 75–76, 109, 151, 162, 163, 164
Amphenol Precision Connector-7, 11
Amplifiers, 90, 99
Analog-to-digital conversion, 208
Anistropic electrically conductive adhesive, 168

APC-7. *See* Amphenol Precision Connector-7
ATE. *See* Automated test equipment
Attenuation, 109
Attenuators, 49, 99, 110, 149
Automated test equipment, 84

Balanced coplanar probe, 67–68, 70
Ball-grid array, 199
Balun, 69–70
Bandpass filter, 46
BCC. *See* Bumped chip carrier
Beryllium-copper, 57–58, 87, 153
BGA. *See* Ball-grid array
Bias, 29, 99, 136, 138, 149–50
Bias cables, 4, 9–10
Bias stability, 45
Bias tee, 10–11, 85
BiCMOS. *See* Bipolar complementary metal-oxide semiconductor
Bipolar complementary metal-oxide semiconductor, 207, 209
Bipolar transistor, 136–37
Blind-mate connector, 85
Block housing, 95–96
Bluetooth, 205

Reproducibility, 48
Resistors, 28, 92, 109, 147
Resonance, measurement, 164
Resonators, 79
    bridged, 168–69
    as verification element, 44–45
Response calibration, 34–35
Response resolution, 178
Return loss, 49, 50, 59, 67, 105, 174,
        179, 183
Reverse-based diode, 147
Reverse reflection calibration, 37
RF. *See* Radio frequency
RFIC. *See* Radio frequency integrated
        circuit
Ridge-trough waveguide, 104
Right-angle thru, 31
Ringing, voltage pulse, 90

Sapphire substrate, 109
Scalar calibration, 44, 93
Scalar de-embedding, 176
Scalar errors, 34
Schottky diode, 99, 147
Schottky noise, 15
Scrub pad, 67
Self-calibration, 24
Self-capacitance, 157
Self-inductance, 165–66, 167
Semirigid cables, 8
Sense connector, 9
Series inductance, 71, 166–67, 183
Series resistance, 120–21, 183
S-G probe. *See* Signal-ground probes
S-G-S. *See* Signal-ground-signal probes
Shielding, 69, 132, 134–36, 147–48
Short, open, and thru method, de-bedding,
        122–23
Shorting transistor, 128–29
Short-open-load, 35, 37, 47, 61, 74, 149
Short-open-load-reciprocal, 37–38, 61
Short-open-load-thru, 34, 35–37, 44, 49,
        61, 73, 76, 78, 97, 106, 109–10
Short standard, 24, 25–26, 38, 41, 42–43,
        47, 50, 174, 183
Shorts (two), open, and thru method,
        de-embedding, 124–26

Signal flow graph, 19–21, 22–23, 24, 46
Signal-ground probes, 59, 68–69
Signal-ground-signal probes, 60, 68
Signal leakage, 46
Signal probes, 59, 67, 136, 138
Silicon, 57, 75, 109, 116–17, 134,
        207, 209
Silicon carbide, 209
Silicon germanium, 207, 209
Single-ended probe, 69–70
Skating, 64–67, 76, 91, 151
Skin depth, 24, 39, 165, 188
Skin effect, 28, 116, 121, 139–40
Skin loss, 72–73
SMA. *See* Subminiature A connector
Smith chart, 13, 25, 26, 28, 37, 44,
        50, 185
SOL. *See* Short-open-load
SOLR. *See* Short-open-load-reciprocal
SOLT. *See* Short-open-load-thru
Source impedance, 145, 146
Source match, 7, 19
S-parameters, 96, 105, 145, 146, 149, 151,
        177, 180, 193–94, 197–98
Spectrum availability, 208
Split block fixture, 95–96
Splitter-combiner balun, 69–70
Spring-loaded contacts, 159
Standard technique, noise
        measurement, 147
Standing-wave ratio, 11
Static system configuration, 18
Step sweep, 7
Stubs, 76, 139–140
Styrofoam, 192
Subminiature A connector, 11, 189
Substrate
    conductive versus insulating, 116–17
    de-embedding, 126–27
    materials, 109, 122, 127, 134, 208–9
    resistivity, 118, 131–32
    temperature changes, 151, 163
    See also Alumina; Gallium arsenide;
        Silicon
Swab cleaning, 67
Sweep modes, 7, 26, 28, 44, 63–64
Systematic error, 15–16

Tantalum nitride resistor, 147
Tapered ground, 102, 103
TC. *See* Thermal coefficient
TCE. *See* Thermal coefficient of expansion
TDR. *See* Time-domain reflectometry
TDT. *See* Time-domain transmission
TEM. *See* Transverse electromagnetic
Temperature, characterizing over, 150–54
Test fixtures
    calibration, 107–12
    components, 95–97, 99
    de-embedding, 171, 174–80
    overview, 2–4, 98, 157
    parasitic effects, 97–98
    passive components, 98–99
    qualities of good, 97
    radio frequency transitions, 99–105
    reference planes, 105–7
Test fixtures, package characterization, 4, 157
    attachment to carrier, 165–70
    body, 159–61
    calibration, 171–74
    carrier, 161–64
    de-embedding, 174–80
    design, 158–61
    RF launchers, 158–59
Test head, 84
Test speed, 2
Test system loss, 79
Test system mismatches, 49
Thermal coefficient, 151
Thermal coefficient of expansion, 153, 163
Thermal expansion, 151–52
Thermal noise, 7, 131, 134
Thermocouple, 154
Thick-film fabrication, 109
Thin-film fabrication, 109, 110
Three-port network, 11
Three-sampler vector network analyzer, 5–6, 22, 34, 40
Thru line, 19, 24, 29–31, 38, 40, 44, 47, 67, 74–75, 76, 93, 146, 168, 172, 179
    de-embedding with, 122–26, 172, 174, 175–76, 180
    package characterization with, 182–83

Thru line discontinuity method, 110
Thru-reflect-attenuate, 40, 110
Thru-reflect-line, 31, 38–41, 44, 73, 76–77, 97, 109–10, 151, 168
Thru-short-delay, 40
Time delay, 32
Time-domain reflectometry, 176–79, 197–98
Time-domain techniques, 93, 128, 175, 176–80
TRA. *See* Thru-reflect-attenuate
Traceability, 47–48
Transfer standard, noise parameters, 147
Transformers, 99–100
Transistor current gain, 118
Transistor finger metal, 128–30
Transistors
    bipolar, 136–37
    field effect, 129, 131, 132, 133, 134, 191, 209
    temperature effect on, 152–54
Transmission lines, 38–41
    discontinuities, 72
    length/width, 40, 77, 90, 97
    loss, 39–40, 117, 164
Transmission mismatch, 7
Transmission normalization, 175
Transmission tracking, 19, 43
Transverse electromagnetic, 77–78, 101, 116
Traveling waves, 60
Triaxial cables, 9–10
TRL. *See* Thru-reflect-line
TSD. *See* Thru-short-delay
Tuners, 146–47, 149
Tungsten probe tips, 57
Tuning stick, 180
Two-port measurement, 19
Two-port network method, 123–24
Two-tier calibration, 107–8

Unbalanced coplanar probe, 68–69, 70
Underfill, 199
Unity-gain amplifier, 9

Varactor diode, 108, 147
Vector calibration, 34–35, 36, 44, 93

Recent Titles in the Artech House Microwave Library

Advanced Techniques in RF Power Amplifier Design, Steve C. Cripps

Behavioral Modeling of Nonlinear RF and Microwave Devices,
Thomas R. Turlington

Computer-Aided Analysis of Nonlinear Microwave Circuits,
Paulo J. C. Rodrigues

Design of FET Frequency Multipliers and Harmonic Oscillators,
Edmar Camargo

Design of RF and Microwave Amplifiers and Oscillators,
Pieter L. D. Abrie

*EMPLAN: Electromagnetic Analysis of Printed Structures in Planarly
Layered Media, Software and User's Manual*, Noyan Kinayman
and M. I. Aksun

Feedforward Linear Power Amplifiers, Nick Pothecary

Generalized Filter Design by Computer Optimization,
Djuradj Budimir

High-Linearity RF Amplifier Design, Peter B. Kenington

Introduction to Microelectromechanical (MEM) Microwave Systems,
Hector J. De Los Santos

Microwave Engineers' Handbook, Two Volumes,
Theodore Saad, editor

*Microwave Filters, Impedance-Matching Networks, and Coupling
Structures*, George L. Matthaei, Leo Young, and E.M.T. Jones

Microwave Materials and Fabrication Techniques, Third Edition,
Thomas S. Laverghetta

Microwave Mixers, Second Edition, Stephen Maas

Microwave Radio Transmission Design Guide, Trevor Manning

Microwaves and Wireless Simplified, Thomas S. Laverghetta

Neural Networks for RF and Microwave Design, Q. J. Zhang and
K. C. Gupta

QMATCH: Lumped-Element Impedance Matching, Software and User's Guide, Pieter L. D. Abrie

RF Design Guide: Systems, Circuits, and Equations, Peter Vizmuller

RF Measurements of Die and Packages, Scott A. Wartenberg

The RF and Microwave Circuit Design Handbook, Stephen A. Maas

RF and Microwave Coupled-Line Circuits, Rajesh Mongia, Inder Bahl, and Prakash Bhartia

RF Power Amplifiers for Wireless Communications, Steve C. Cripps

RF Systems, Components, and Circuits Handbook, Ferril Losee

TRAVIS 2.0: Transmission Line Visualization Software and User's Guide, Version 2.0, Robert G. Kaires and Barton T. Hickman

Understanding Microwave Heating Cavities, Tse V. Chow Ting Chan and Howard C. Reader

For further information on these and other Artech House titles, including previously considered out-of-print books now available through our In-Print-Forever® (IPF®) program, contact:

Artech House
685 Canton Street
Norwood, MA 02062
Phone: 781-769-9750
Fax: 781-769-6334
e-mail: artech@artechhouse.com

Artech House
46 Gillingham Street
London SW1V 1AH UK
Phone: +44 (0)20 7596-8750
Fax: +44 (0)20 7630 0166
e-mail: artech-uk@artechhouse.com

Find us on the World Wide Web at:
www.artechhouse.com